William Arthur Clark

Physiology

A Manual of 1000 Questions and Answers Systematically Arranged

William Arthur Clark

Physiology
A Manual of 1000 Questions and Answers Systematically Arranged

ISBN/EAN: 9783744670067

Printed in Europe, USA, Canada, Australia, Japan

Cover: Foto ©berggeist007 / pixelio.de

More available books at **www.hansebooks.com**

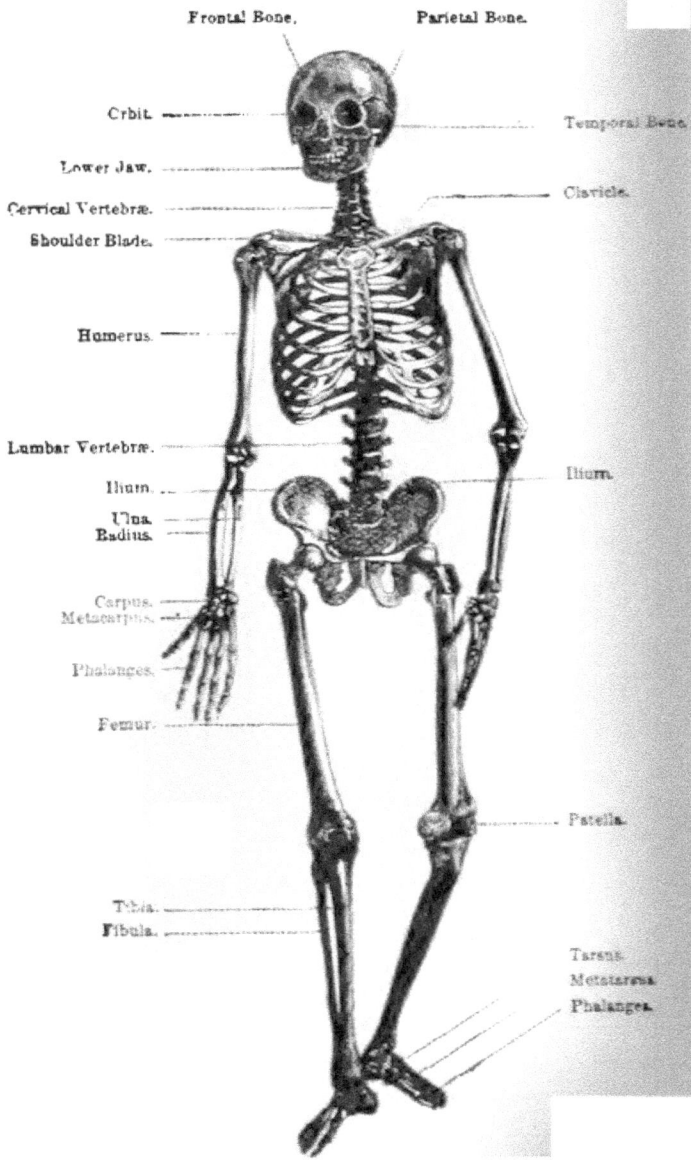

Frontal Bone.   Parietal Bone.

Orbit.

Temporal Bone.

Lower Jaw.

Clavicle.

Cervical Vertebræ.

Shoulder Blade.

Humerus.

Lumbar Vertebræ.

Ilium.

Ilium.

Ulna.
Radius.

Carpus.
Metacarpus.

Phalanges.

Femur.

Patella.

Tibia.
Fibula.

Tarsus.
Metatarsus.
Phalanges.

THE

ANALYTIC SERIES

OF

TEACHERS' AIDS.

PHYSIOLOGY:

A MANUAL OF 1000 QUESTIONS AND ANSWERS SYSTEMATICALLY ARRANGED,

CONTAINING A

FULL TREATMENT OF THE PHYSIOLOGICAL EFFECTS OF ALCOHOL AND
NARCOTICS, WITH A COMPLETE ANALYTIC OUTLINE OF
THE SUBJECT, AND NOTES ON TEACHING.

AN AID IN TEACHING AND IN PREPARING FOR EXAMINATION.

By W. A. CLARK,
Instructor in College of Science, National Normal University.

C. K. HAMILTON & CO.
LEBANON, OHIO.

PREFACE.

THIS little manual finds the warrant for its existence in the continued demand for "Question Books" on the Common School Branches. Its aim is to furnish teachers the proper materials for a rapid but comprehensive review of the subject. It may also be used as a basis of class work, furnishing an orderly arrangement of topics for investigation and discussion. The matter has been carefully selected; and it is believed it will be found to cover, with some degree of completeness, the entire field. The Analytic Outline, which is the skeleton of the entire body of questions, will greatly assist the memory by giving system and definiteness to the study. It is as logical as is consistent with the design of the book. The illustrations will aid in understanding the answers to many questions. The Didactic Notes, necessarily very brief, contain suggestions of methods of teaching the various subjects, and indicate sources of materials for illustration and experiment. It is hoped that they may aid the progressive teacher in his efforts to do pleasant and profitable work.

CONTENTS.

GENERAL DIDACTIC NOTE.

THE successful teacher of Physiology must be an enthusiastic student, a leader of enquiring minds seeking a knowledge of their own clay tenements, rather than a formal instructor in scientific truth collected and systematized by others. He must be able by the power of his imagination to transform the skeleton from a mere collection of bones into a beautifully contrived mechanism for protection, support, and locomotion; he must see in the heart not a hollow muscle merely, but a wonderful little engine, pumping the life fluid into every cranny of the "living walls;" and to him the eye must be not only a finely constructed optical instrument, but a veritable "window of the soul." His success depends upon his power of idealizing.

Physiology is pre-eminently suited to investigation by outlining. A General Outline should first be presented to the class, or rather developed from the pupils themselves by skillful questioning, and then drilled upon thoroughly day after day until each pupil can give it orally or place it on the blackboard from memory. Various classifications and forms of the general outline may be used, to suit the grade of the class and the views of the teacher. The Special Outline for each separate division of the subject should be developed by the pupils in their own preparation of the lesson, and then discussed and revised in the recitation.

Important points should be made prominent, and many minor matters which are of value only to specialists should be entirely omitted. Beware of trying to teach too much. Give the instruction in Hygiene the first place, remembering, however, that all rational views on hygienic matters are conditioned on a knowledge of the structure and functions of the various organs of the body.

The use of materials obtained from the butcher's shop and other sources seems so necessary to good teaching as to render superfluous any recommendation here; but it is well known that many good teachers fail of their highest success by neglecting to use such materials as can be readily obtained in the neighborhood of any school. Make the phenomena real by presenting them before the class in actual existence as well as in descriptive words. *Use the blackboard!*

The order of subjects adopted in this book is well suited to advanced classes; but for more elementary instruction the simplest and most direct method of developing the science is by beginning with the study of the articulated skeleton.

The instruction on Alcohol and Narcotics should all tend toward the prevention of the use of intoxicating drinks and tobacco; but good hygienic instruction here, as elsewhere, is based upon a knowledge of the nature of the poisons and their effects upon man in his three-fold nature, physical, intellectual, and moral. No theories of individual responsibility and personal liberty should keep the teacher from doing his whole duty

GENERAL OUTLINE.

SCOPE AND DIVISIONS OF THE SCIENCE.

Didactic Note. — In commencing the study of any science, the pupil should have a clear conception of the general character of the subject-matter, the divisions of the science, and its relations to the other departments of human knowledge. With these facts well fixed in the mind, he has a frame-work on which to build his future acquisitions. A simple understanding of the nature of the subject is of much greater importance than the mere memorizing of technical terms. The teacher should encourage the pupil to make his statements in his own language, and to illustrate them with examples from his own stock of knowledge. Be satisfied to build slowly. The terms given below should be taught incidentally throughout the study of the subject—not as mere words, but as embodying the great generalizations of the science. No attempt to outline this division has been made, since it is necessarily incomplete.

1. *Define Anatomy.*
Anatomy is the science of the structure of organized bodies.
2. *Define Physiology.*
Physiology is the science of the functions of organized bodies.
3. *Define Hygiene.*
Hygiene is the science of health.

9

4. What does the term Physiology commonly include?

The term Physiology, as it is used in elementary text-books on the subject, includes Anatomy, Physiology and Hygiene.

5. Define Human Anatomy.

Human Anatomy is the science of the structure of the human body.

6. Define Comparative Anatomy.

Comparative Anatomy is the comparative study of the same organs in man and the inferior animals.

7. Define Histology.

Histology, or General Anatomy, is a study of the microscopic structure of tissues.

8. Define Descriptive Anatomy.

Descriptive Anatomy is a study of the forms and relations of the various organs of the body.

9. Define Pathology.

Pathology is the science of diseases.

10. Define Biology.

Biology is the science of life.

11. Define Osteology.

Osteology is that division of Physiology which treats of the bones.

12. Define Arthrology.

Arthrology is that division of Physiology which treats of the joints.

13. Define Myology.

Myology is that division of Physiology which treats of the muscles.

14. Define Splanchnology.

Splanchnology is that division of Physiology which treats of the digestive apparatus.

15. Define Angiology.

Angiology is that division of Physiology which treats of the circulatory apparatus.

16. Define Pneumology.

Pneumology is that division of Physiology which treats of the respiratory apparatus.

17. *Define Ichorology.*

Ichorology is that division of Physiology which treats of the absorptive apparatus.

18. *Define Neurology.*

Neurology is that division of Physiology which treats of the nervous apparatus.

CELLS.

Didactic Note.— The minute structure of the human body, which is revealed only by the microscope, is usually too much neglected in the teaching of elementary Physiology. This division of the subject presents no inherent difficulty, and possesses many attractions for pupils. A clear idea of the *cell* is the prime essential. This any teacher can secure by a proper use of the blackboard. The amœba is an excellent example of an individual cell. The nature of life and death as states of organized being should be carefully taught by numerous examples.

SPECIAL OUTLINE.

1^1 Cells.   [See General Outline, page 7.]

1^2 General description.

1^3 The ideal cell.

2^3 Form.

3^3 Size.

2^2 Structure.

1^3 Cell-wall.

2^3 Cell contents.

1^4 Protoplasm.

2^4 Granules.

3^4 Nucleus.

4^4 Nucleolus.

3^2 Classification.

1^3 Immersed in a liquid.

2^3 Imbedded in a solid.

3^3 On a free surface.

4^2 The Amœba.

5^2 Cell development.

1^3 Growth.

2^3 Reproduction.

1^4 Segmentation.

2^4 Gemmation.

6^2 Vitality.

19. *What are cells?*

Cells are minute masses of simple formative material. **They** are the anatomical units of all organized matter.

20. *What is the ideal cell?*

The ideal cell is a small sac, the Cell-wall, enclosing a granular mass of Protoplasm in which there is imbedded a vesicle, the Nucleus, containing a still smaller body, the Nucleolus.

21. *Where are ideal cells found?*

Ideal cells are found in the blood, where **they are called** *white corpuscles.*

22. *Do all cells conform to this perfect type?*

Cells frequently want the enclosing membrane; and the **little** spot within the vesicle is sometimes not distinguishable. **Some** histologists apply the term cell to minute masses of **structure**-less matter **in** which no vesicle is discernible.

23. *What is a Nucleated Cell?*

A nucleated cell is one containing **a small vesicle, or kernel,** which is the center of cell activity.

24. *What forms do cells assume?*

The ideal cell is spherical in form ; but cells assume in different positions and under different circumstances a great variety of forms—cylindrical, conical, disk-like, etc. A peculiar **form** is the ciliated cell.

25. *Describe Ciliated Cells.*

Ciliated cells are usually conical in form, and are placed on the walls of cavities with their bases, from which *cilia* or hair-like bodies project, turned inward towards the cavity

26. *What is the size of cells?*

Cells vary in size from $\frac{1}{120}$ to $\frac{1}{10000}$ of an inch **in diameter, the** average size being about $\frac{1}{3000}$ of an inch.

27. *What is the Cell-wall?*

The cell-wall is **the** enclosing membrane—frequently only the inseparable outer layer of the cell-body.

28. *What is the Cell-body?*

The cell-body is the semi-fluid mass contained within the cell-wall.

29. *What is Protoplasm?*

Protoplasm is the simplest formative material **of animal and vegetable** structures, having no distinct tissues. **It composes** the cell-body.

30. *What are Granules?*

Granules are minute specks floating in the protoplasm.

31. *What is the Nucleus?*

The nucleus is a small vesicle usually attached to the inner ..de of the cell-wall and containing a semi-transparent mass.

32. *What is the Nucleolus?*

The nucleolus is a minute speck, or granule, within the nucleus.

33. *Into what three classes are cells divided with respect to location?*

Cells are divided into three classes, as follows: 1st, those immersed in fluids; 2nd, those imbedded in solids; and 3rd, those on free surfaces.

34. *What cells belong to the first group?*

The corpuscles of the blood and lymph belong to the first group, or class.

35. *What cells belong to the second group?*

Those found in bone, cartilage, and other connective tissues.

36. *What cells belong to the third group?*

The epithelium cells of the skin and mucous membrane.

37. *What is the Amœba?*

The Amœba is a microscopic animal consisting of a single cell.

38. *What vital properties does the Amœba possess?*

It possesses the powers of assimilation, reproduction, irritability, and contractility. It can change its form at will, and can move about from place to place.

39. *What two processes are denoted by the term development, as applied to cells?*

The growth of existing cells and the producing of new ones.

40. *How does a cell grow?*

A cell grows by incorporating new material among its constituent molecules. This growth by interstitial deposit is called *intussusception*, as opposed to *accretion*, or formation by adding layers to the outside, as in the mineral world.

41. *By what two principal modes is the reproduction of cells accomplished?*

The reproduction of cells is accomplished principally by *segmentation* and *gemmation*.

42. *Describe the process of Segmentation.*

In Segmentation the "mother-cell" becomes gradually constricted in its center, like an hour-glass, and finally separates into two cells, each containing a part of the divided nucleus.

43. *Describe the process of Gemmation.*

In Gemmation a bud, or little elevation, first projects from the cell mass, and subsequently separates by constriction at its base.

44. *Define growth.*

The term growth denotes increase in the size of a structure without change in the nature of its fabric or in the functions it discharges.

45. *How does the human body grow?*

The human body grows by the multiplication of its cells.

46. *What is Vitality?*

Vitality is the life force of animals and plants. It is the force which, controlling chemical and physical forces, gives form to organized matter. A body is said to live when this force is operative in it, and to die when this force departs from it.

47. *Define life as a condition.*

Life is that condition of organized being in which it is capable of performing all its natural functions.

48. *Define death.*

Death is that condition of organized being in which it is incapacitated for performing its various functions. Death may be partial and gradual, organ after organ losing its power to act until total incapacity results.

CHEMICAL ELEMENTS AND PROXIMATE PRINCIPLES.

Didactic Note.—The nature of a chemical element can be taught to those who have not studied chemistry by a few simple examples, as gold, iron, etc. A clear understanding of the terms *chemical compound* and *proximate principle* can be secured in the same way. The extent to which the subdivision of organic proximate principles is carried should depend upon the grade of the pupils ; but the teacher should, at least, carefully make the distinction between nitrogenous and non-nitrogenous principles.

SPECIAL OUTLINE.

2^1 Chemical Elements. [See General Outline, page 7.]
  1^2 Nature.
  2^2 Number.
  3^2 Enumeration.
  4^2 Combination.
3^1 Proximate Principles.
  1^2 Nature.
  2^2 Classification.
    1^3 Inorganic.
      1^4 Water, etc.

    2^3 Organic.
      1^4 Nitrogenous.
        1^5 Proteids.
        2^5 Peptones.
        3^5 Albuminoids.
        4^5 Coloring matters.
      2^4 Non-nitrogenous.
        1^5 Hydrocarbons.
        2^5 Carbohydrates.

49. *What are Chemical Elements?*

Chemical Elements are simple substances which are recognized by chemists as ultimate forms of matter. About seventy elements are now known.

50. *What Chemical Elements are found in the human body?*

The following *sixteen* elements are found in the human body: Carbon, Hydrogen, Nitrogen, Oxygen, Sulphur, Phosphorus, Chlorine, Fluorine, Sodium, Potassium, Calcium, Magnesium, Silicon, Lithium, Iron, and Manganese.

51. *Do these elements exist in a free state in the body?*

The chemical elements exist in the various tissues of the human body in a combined state. A small amount of free oxygen is found in the blood; a small amount of nitrogen, in the lungs; etc.

52. *What are Proximate Principles?*

The proximate principles of the human body are simple substances which cannot be separated into simpler components without the aid of chemical analysis.

53. *Into what two classes are proximate principles divided?*

The proximate principles of the body are divided into two great classes: Inorganic and Organic.

54. *Name the most important inorganic proximate principles.*

Water, Common Salt, Potassium Chloride and Calcium Phosphate.

55. *What is the quantity of water in the human body?*

Water constitutes about two-thirds, by weight, of the body, and is found in all its tissues.

56. *Where is* **common salt** *(Sodium Chloride) found in the human body?*

Common salt is found in all **the** tissues of the body.

57. *In what tissues is Potassium Chlorate found?*

Potassium Chlorate is found in the blood, muscles, nerves, and in most of the fluids of the body.

58. *Where is Calcium Phosphate found?*

Calcium Phosphate is found in all the tissues, but chiefly in the bones and teeth. It constitutes more than half of the material of the bones.

59. *Into what two classes are the organic proximate principles divided?*

Nitrogenous and Non-nitrogenous principles.

60. *What are the characteristics of the nitrogenous compounds?*

Since these compounds contain nitrogen, they are very unstable and are constantly changing. They give to tissues and fluids their vital properties.

61. *Into what classes are they grouped?*

Proteids, Peptones, Albuminoids and Coloring Matters.

62. *What are Proteids?*

Proteids are very complex chemical compounds, composing the principal solids of the muscular, nervous and glandular tissues, and of the plasma of the blood. The white of an egg belongs to this class.

63. *What are Peptones?*

Peptones are compounds resulting from the action of the fluids of the alimentary canal upon proteids.

64. *What are Albuminoids?*

Albuminoids are compounds closely related to the proteids, but differing from them in the fact that they contain no sulphur.

65. *What are Nitrogenous Coloring Matters?*

They are a non-related class of substances, which give color to the fluids of the body. To this group belong *Hæmatin* and

Hæmoglobine in the blood, and *Bilirubin* and *Biliverdin* in the bile.

66. *What are the Non-nitrogenous Compounds?*

These substances are composed of Carbon, Hydrogen and Oxygen. They contain no Nitrogen, and are, therefore, more stable than the nitrogenous group.

67. *Into what two classes are the non-nitrogenous principles divided?*

Hydrocarbons, or the fats of the body; and Carbohydrates, or starch and various forms of sugar.

TISSUES.

Didactic Note.—No part of Physiology will better repay the effort of close study than the Tissues. Their beauty of structure, as revealed by the microscope and pictured in many school physiologies, can be shown in rapid blackboard sketches. Let no teacher plead inability to use the crayon in this way. It is not the work of an *artist*, but of a *teacher;* and facility here as elsewhere is the result of practice. Get a bone from the butcher's shop, and saw it across to show bone structure; burn it in the fire to destroy the animal matter; and mascerate it in diluted sulphuric acid to dissolve out the mineral matter. "Lean meat" from the same source will show the muscular tissues, etc. Remind the pupils constantly of facts already observed by them at their homes: as the breaking down of the structure of boiled beef into minute fibres, or *fasciculi;* the separating of a piece of boiled tendon into slender, thread-like fibres; the white, crackling areolar tissue seen immediately beneath the skin in butchering a beef; etc.

SPECIAL OUTLINE.

4^1 Tissues.   [See General Outline, page 7.]

1^2 Differentiation.

2^2 Classification.

1^3 Supporting.

1^4 Sclerous.

1^5 Chemical Composition.

2^5 Microscopic Structure.

3^5 Ossification.

2^4 Fibrous Connective.

1^5 Inelastic.

2^5 Elastic.

3^5 Areolar.

3^4 Cartilaginous.

1^5 Hyaline.

2^5 Elastic.

3^5 Fibro-cartilaginous.

2^3 Contractile: Muscular.

1^4 Involuntary.

1^5 Structure.

2^5 Functions.

2^4 Voluntary.

1^5 Structure.

2^5 Functions.

3^3 Storage: Adipose.

1^4 Structure.

2^4 Where found.

4^3 Conductive: Nervous.

1^4 Cells.

2^4 Fibres.

1^5 Structure.

2^5 Kinds.

1^6 White.

2^6 Gray.

68. *What are Tissues?*

Tissues are the simple organic structures composing the human body. They are building materials, bearing the same relation to the body as a whole that the wood, paint, glass, slate, and nails do to the house.

69. *How are the different kinds of tissues formed?*

The various forms of tissue are developed from simple cells by a process known as the *differentiation of the tissues*, in which cells at first similar become by slight modifications gradually changed into forms fitted for special functions.

70. *Enumerate the tissues of the human body.*

Sclerous, Fibrous Connective, Cartilaginous, Muscular, Adipose, and Nervous.

71. *How are the tissues classified?*

Various schemes of classification have been adopted by different writers. Perhaps the most satisfactory division is into four groups, as follows: Supporting, including the Sclerous, Fibrous

Connective, and Cartilaginous; Contractile, or Muscular; Storage, or Adipose; and Conductive, or Nervous.

72. *What is Sclerous tissue?*

Sclerous tissue is a porous, mineral structure composing the bones and teeth. The word sclerous means hard.

73. *Give the chemical composition of bones.*

The following from Berzelius, as quoted in *Gray's Anatomy*, is reliable:

Organic matter (Gelatin and blood-vessels),	33.30
Inorganic, or earthy matter—	
Phosphate of Lime,	51.04
Carbonate of Lime,	11.30
Fluoride of Calcium,	2.00
Phosphate of Magnesia,	1.16
Soda and Common Salt,	1.20
	100.00

74. *What are the two forms of sclerous tissue?*

Sclerous tissue is of two distinct varieties: Cancellous, a bony network of minute spiculæ found in the irregular bones and in the enlarged ends of the long bones; and Compact, a denser tissue consisting of minute plates of bone arranged in what are called *Haversian systems*, and found in the shafts of long bones and in a hard surface layer of the irregular ones.

75. *Describe an Haversian System?*

An Haversian System consists of a central *Haversian Canal* surrounded by concentric layers of bony *Lamellæ*, between which are rows of *Lacunæ* joined together by radiating *Canaliculi*.

76. *What are Haversian Canals?*

Haversian Canals are minute tubes, about $\frac{1}{500}$ of an inch in diameter, running through the substance of bones and containing nutrient vessels. The entrances to the larger ones from the surfaces of bones and the medullary cavities are called *Nutrient Foramina*.

77. *What are Lamellæ?*

Lamellæ are little plates or scales of bone arranged in concentric rings around the Haversian canals.

78. *What are Lacunæ?*

Lacunæ are small, irregularly-shaped cavities between the Lamellæ.

79. *What are Canaliculi?*

Canaliculi are minute tubes radiating from Haversian canals and joining the rings of lacunæ into systems.

80. *What are Bone Cells?*

Lacunæ, with their contents, are called Bone Cells, or *Bone Corpuscles.* The cells which are centers of intramembranous ossification are called *Osteoblasts.*

81. *What are Ultimate Granules?*

Ultimate Granules are minute particles of mineral matter about $\frac{1}{6000}$ of an inch in diameter, out of which the bone structure is built.

82. *What is Ossification?*

Ossification is the formation of bone by the depositing of mineral matter in cartilaginous or membranous structures.

83. *What are the two kinds of ossification?*

Intracartilaginous and Intramembranous.

84. *Describe Intracartilaginous Ossification.*

There are four distinct stages in the process of intracartilaginous ossification, as follows: 1st, a *center of ossification* is formed by the enlarging and arranging in rows of the cartilage cells; 2nd, calcareous matter is deposited in the spaces between the cells; 3rd, processes of blood-vessels and cells burrow into this substance from without; and 4th, little plates of bone substance arrange themselves in concentric rings about these burrows, or tubes, and form Haversian systems.

85. *Describe Intramembranous Ossification?*

In intramembranous ossification the center of ossification is a granular corpuscle, from which spiculæ of bone substance radiate in all directions, much like the formation of ice on the surface of water. In other respects the process is similar to that in intracartilaginous ossification.

86. *What are Fibrous Connective Tissues?*

Fibrous Connective Tissues are a class of textures whose common function is to bind together and support the various organs of the body.

87. *Of what are these tissues composed?*

The fibrous connective tissues are composed of a basement substance in which are imbedded cells, and through which run two kinds of fibres, *elastic* and *inelastic.*

88. *What are the three varieties of fibrous connective tissues?*

White Fibrous Tissue, Yellow Elastic Tissue, and Areolar Tissue.

89. *Describe White Fibrous Tissue.*

White Fibrous Tissue is composed almost wholly of inelastic fibres, is white in appearance, and is very strong. It constitutes the ligaments for binding together the bones, the tendons for joining muscles to bones and other structures, and the investing membranes for the protection of various parts, as the coverings of bones and muscles.

90. *Describe Yellow Fibrous Connective Tissue.*

Yellow Fibrous Connective Tissue consists largely of yellow elastic fibres, which give it a yellowish color and render it elastic. It is found in the vocal cords, the inner coats of the arteries, etc.

91. *Describe Areolar Tissue.*

Areolar Tissue consists of open meshes of mingled elastic and inelastic fibres. It is found in nearly every part of the body, binding together and supporting the various organs.

92. *What are Basement Membranes?*

Basement Membranes are a special form of connective tissue found in cell-walls, supporting membranes, etc. They are the simplest of all tissues.

93. *What is Cartilaginous Tissue?*

Cartilaginous Tissue is a dense, tough substance found in the joints and elsewhere, and commonly called " gristle."

94. *Into what classes are cartilages divided?*

Cartilages are divided into three classes: Hyaline, Elastic, and Fibro-Cartilage.

95. *Describe Hyaline Cartilage.*

Hyaline Cartilage is a pale, bluish-white structure, flexible and elastic, found in the joints, where it is called *Articular Cartilage*, and at the juncture of the ribs and sternum, where it is called *Costal Cartilage.* The *Temporary Cartilage*, from which the bones are formed, is of this variety.

96. *Describe Elastic Cartilage.*

Elastic Cartilage is yellowish in appearance, and differs from the *true*, or hyaline cartilage in the presence of fibres of yellow elastic connective tissue interlacing through its structure. It is found in the external ear, the walls of the nose, the epiglottis, etc.

97. *Describe Fibro-Cartilage.*

Fibro-Cartilage consists of true cartilage substance with interlacing threads of white fibrous connective tissue. It is found in the mechanism of certain joints which are subject to frequent movement and violent concussion, as the knee, the wrist, etc.

98. *What is Muscular Tissue?*

Muscular Tissue is a highly contractile texture of fibrous structure composing the muscles.

99. *What are the two varieties of muscular fibres?*

Involuntary and Voluntary.

100. *Describe the structure of Involuntary Muscular Fibres.*

Involuntary Muscular Fibres are spindle-shaped contractile fibres, or elongated cells, showing under the microscope a smooth, plain structure, hence called "unstriped muscular fibres."

101. *Where are involuntary, or unstriped muscular fibres found?*

Involuntary muscular fibres are found in the muscles of the walls of the viscera and other organs not under the immediate control of the will.

102. *Describe the structure of Voluntary Muscular Fibres.*

Voluntary or "Striped" Muscular Fibres are thread-like bodies about $\frac{1}{400}$ of an inch in diameter, marked by transverse bands, and each enclosed in a delicate sheath, called the *Sarcolemma.* According to some authorities, these fibres are composed of still finer threads, called *Fibrillæ.*

103. *Where are voluntary muscular fibres found?*

Voluntary muscular fibres compose the muscles which are attached to the bones, and which are, therefore, used under the control of the will in the ordinary movements of the body.

104. *Describe the structure of a voluntary muscle.*

A voluntary muscle is composed of little bundles of striped fibres, called *Fasciculi,* bound together by a closely investing membrane called the *Perimyssium.*

105. *What is Muscle Juice?*

Muscle Juice is a colorless fluid permeating the muscular structure.

106. *How do muscles contract?*

Muscles contract by the shortening of their fibres. The fibre does not decrease in volume, but simply changes its form, becoming shorter and thicker.

107. *What is Adipose Tissue?*

Adipose Tissue consists of little vesicles containing fatty matter bound together by connective tissue.

108. *Where is adipose tissue found?*

Adipose tissue is found in almost every part of the body, distributed through the meshes of areolar tissue. It is a store-house for heat-forming materials.

109. *What is Nervous Tissue?*

Nervous Tissue is the texture of the sensory apparatus of the body, and is composed of cells interlaced with fibres and bound together by connective tissue.

110. *Where are Nerve Cells chiefly found?*

Nerve cells are collected in masses, intermingled with fibres, in the brain and other ganglionic centers, constituting the *gray nervous substance.*

111. *What are the two kinds of Nerve Fibres?*

Nerve Fibres are White, or Medullated, and Gray, or Non-medullated.

112. *Describe the structure of White Nerve Fibres.*

White Nerve Fibres consist of a central core, or *Axis Cylinder*, imbedded in a fatty substance, called the *White Substance of Schwann*, and the whole surrounded by a delicate membrane called the *Neurilemma.*

113. *Describe the structure of Gray Nerve Fibres.*

Gray Nerve Fibres are bundles of minute, finely-striated fibrillæ, enclosed in a delicate sheath.

114. *Where are white nerve fibres chiefly found?*

White nerve fibres constitute the white part of the brain and spinal cord and the greater part of the cerebro-spinal nerves.

115. *Where are the gray nerve fibres chiefly found?*

Gray nerve fibres compose the nerves of the sympathetic system, and are also found in the nerve centers and in some of the cerebro-spinal nerves.

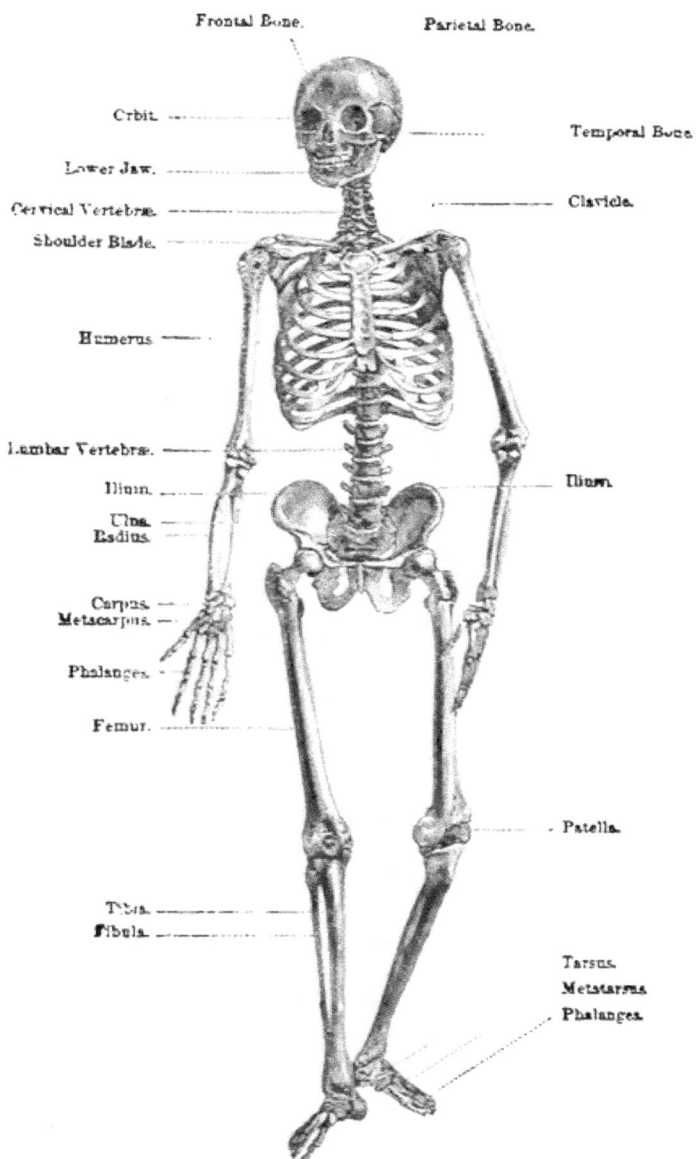

Frontal Bone.

Parietal Bone.

Orbit.

Temporal Bone.

Lower Jaw.

Cervical Vertebræ.

Clavicle.

Shoulder Blade.

Humerus.

Lumbar Vertebræ.

Ilium.

Ilium.

Ulna.

Radius.

Carpus.

Metacarpus.

Phalanges.

Femur.

Patella.

Tibia.

Fibula.

Tarsus.

Metatarsus.

Phalanges.

OSSEOUS SYSTEM.

Didactic Note.—Where teaching is done without the aid of an artificial human skeleton, as it is in most schools, the teacher must use such material as can be obtained, remembering the value of seeing and handling that concerning which we wish to know. Pupils, if encouraged to do so, will soon make a good collection of *clean* bones of animals, by the aid of which the forms, processes, depressions, etc., of bones can be delightfully taught. Blackboard drawings, more or less conventional, are an important means to the same end. A full outline of the human skeleton should be placed upon the board, first by the teacher and then by the pupils from memory, and drilled upon day after day. The lessons on the functions of bones, accidents, causes of deformities, etc., are specially important. Refer constantly to what the pupils already know concerning themselves.

SPECIAL OUTLINE.

1^1 Osseous System.   [See General Outline, page 7.]
$\quad 1^5$ Bones.
$\quad\quad 1^6$ Mechanical Structure.
$\quad\quad\quad 1^7$ External Examination.
$\quad\quad\quad\quad 1^8$ Eminences: Processes.
$\quad\quad\quad\quad\quad 1^9$ Tuberosities, etc.
$\quad\quad\quad\quad 2^8$ Depressions.
$\quad\quad\quad\quad\quad 1^9$ Fossæ, etc.
$\quad\quad\quad 2^7$ Transverse Section.
$\quad\quad\quad\quad 1^8$ Periosteum, etc.
$\quad\quad\quad 3^7$ Longitudinal Section.
$\quad\quad\quad\quad 1^8$ Shaft, etc.
$\quad\quad 2^6$ Classification.
$\quad\quad\quad 1^7$ Long.
$\quad\quad\quad 2^7$ Short.
$\quad\quad\quad 3^7$ Flat.
$\quad\quad\quad 4^7$ Irregular.
$\quad 2^5$ Skeleton.
$\quad\quad 1^6$ Architecture of skeleton.
$\quad\quad 2^6$ Number of bones.
$\quad\quad 3^6$ Adventitious bones.
$\quad\quad 4^6$ Analysis of skeleton.
$\quad\quad\quad 1^7$ Head.
$\quad\quad\quad\quad 1^8$ Cranium.
$\quad\quad\quad\quad\quad 1^9$ Occipital, etc.
$\quad\quad\quad\quad 2^8$ Face.
$\quad\quad\quad\quad\quad 1^9$ Malar, etc.

3^2 Ear.
    1^7 Maleus, etc.
2^7 Trunk.
    1^8 **Spinal Column.**
      1^9 **Cervical Vertebræ, etc.**
    2^8 Ribs.
      1^9 True, etc.
    3^8 Sternum.
      1^9 Manubrium, etc.
    4^8 Os Hyoides.
    5^8 Pelvis.
      1^9 Innominata, etc.
3^7 Extremities.
    1^8 Upper.
      1^9 Shoulder.

1^{10} Scapula, **etc.**
2^9 Arm : Humerus.
3^9 Forearm.
    1^{10} Radius, etc.
4^9 Hand.
    1^{10} Carpus, etc.
2^8 Lower.
    1^9 Thigh : Femur.
    2^9 **Leg.**
      1^{10} **Tibia, etc.**
    3^9 **Foot.**
      1^{11} **Tarsus, etc.**
3^3 Functions of bones.
4^3 Hygiene.

---

116. *What is an Organ?*

An Organ is a distinct part of the body, designed to perform a particular function, as the heart to propel the blood.

117. *What is the function of an organ?*

The function of an organ is its use or office, as *sight* is the function of the *eye*.

118. *What is a System?*

A System is an assemblage of organs, usually composed of the same tissues, and having similar or related functions.

119. *What is an Apparatus?*

An Apparatus is a collection of systems and organs accomplishing together a common end.

120. *Into what two classes may the apparatuses of the body be divided?*

Apparatuses of Internal Mechanism and of External Relations.

121. *What two apparatuses constitute the first class?*

The apparatuses of internal mechanism are the Motor and the Nutritive.

122. *What is the Motor Apparatus?*

The Motor Apparatus is a collection of systems having for their common end the various movements of the body.

123. *What three Systems compose the Motor Apparatus?*

Osseous, Muscular, and Articulatory.

124. *What is the Osseous System?*

The term Osseous System denotes collectively the entire bony structure of the body.

125. *What is a Bone?*

A Bone is one of the hard parts which compose the *Skeleton,* or frame-work of the body.

126. *In what three ways may the mechanical structure of bones be studied?*

By External Examination, by Transverse Section, and by Longitudinal Section.

127. *What does an External Examination show?*

Eminences and Depressions.

128. *Into what principal classes, with respect to form, are Eminences, or "Processes" as they are called, divided?*

Processes, when the elevations are broad, rough and uneven, are called *Tuberosities;* when small and rough, *Tubercles;* when sharp, slender and pointed, *Spines;* and when narrow, rough ridges, *Lines.*

129. *Into what classes, with respect to form, are the depressions on the surfaces of bones divided?*

When the depression has an entrance larger than the bottom, it is called a *Fossæ;* when it is a long, narrow cleft, it is called a *Fissure;* when it is an indentation in the edge of a bone, it is called a *Notch;* when it is a long cleft or channel wider than a fissure, it is called a *Groove;* and when it is an opening piercing through a bone, it is called a *Foramen.*

130. *Into what two classes, with respect to position, are eminences and depressions alike divided?*

Articular when near a joint, and Non-Articular when not near a joint.

131. *What does a transverse section of a long bone show?*

A membranous covering, called the *Periosteum;* a compact outer layer of bone tissue; a more porous inner body; and a central canal, containing a fatty substance, called the *Medulla,* or "marrow."

132. *What does a longitudinal section of a long bone show?*

A hollow, cylindrical *Shaft* of dense bone tissue, with enlarged *Extremities* of more porous matter.

133. *Why are the extremities of the long bones enlarged?*

To render them more spongy and yielding to sudden jars; to offer better articulatory surfaces; and to afford more room for the attachment of muscles.

134. *What does a section of a bone of the cranium show?*

Two compact laminæ, or plates, called the *Outer and Inner Tables,* having between them a more porous bone structure, called the *Diploic Tissue.*

135. *Into what four classes are bones divided, with respect to form?*

Long, Short, Flat, and Irregular.

136. *Where are the Long Bones found?*

Long bones are found in the extremities, where they serve as levers in producing the various movements of the body.

137. *Where are Short Bones found?*

Short bones are found in the wrist and ankle, where strength and compactness with slight motion are required.

138. *Where are Flat Bones found?*

Flat bones are found where protecting walls for vital organs are necessary, as in the skull, and where broad surfaces for muscular attachment are required, as in the shoulder-blade.

139. *Where are Irregular Bones found?*

Irregular bones are found where a variety of uses are to be subserved, as in the Vertebræ, where strength, protection, and large muscular attachment are all required.

140. *What is the Skeleton?*

The Skeleton is the bony frame-work of the body.

141. *What other supporting frame-works has the human body?*

Besides the *Endoskeleton,* or bony frame-work, the human body has two other supporting structures: the skin, called the *Exoskeleton,* and the connective tissue, called the *Mezzoskeleton.*

142. *What is an Artificial Skeleton?*

An Artificial Skeleton is the rearticulated bones of the dissected body, joined together by wires and springs.

143. *Describe the architecture of the skeleton.*

The human skeleton consists of a central axis, along which are found three enlargements, or protecting cavities, and from which branch two pairs of limbs.

144. *How many bones are there in the human skeleton?*

The skeleton is composed of 206 bones, including the bones of the ear and excluding various minor sessamoid bones.

145. *Why does the number of bones vary at different ages?*

Several bones which in youth are separate become united by ossification in old age, as in the Sternum, the Sacrum, etc.

146. *What are Sessamoid Bones?*

Sessamoid bones are small bones found near the joints. The Patellæ are the largest and most important sessamoid bones.

147. *What are Wormian Bones?*

Wormian bones are small bones frequently found in the sutures of the cranium.

148. *Into what three divisions are the bones of the articulated skeleton usually grouped?*

Bones of the Head, Trunk, and Extremities.

149. *What are the three divisions of the bones of the head, or skull?*

Cranium, Face, and Ears.

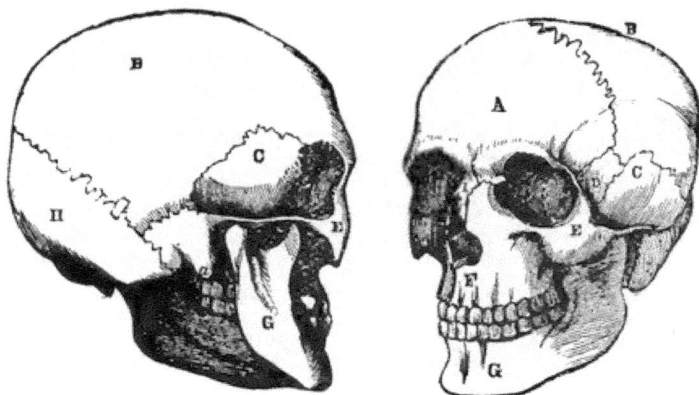

BONES OF THE HEAD.

A, Frontal. *B*, Parietal. *C*, Temporal. *D*, Sphenoid. *E*, Malar. *F*, Superior Maxillary. *G*, Inferior Maxillary. *H*, Occipital. *I*, Nasal.

150. *Enumerate the bones of the Cranium?*

There are eight bones of the cranium: 1 Occipital, 2 Parietal, 2 Temporal, 1 Frontal, 1 Sphenoid, and 1 Ethmoid.

151. *Enumerate the bones of the Face.*

There are fourteen bones in the face : 2 Malar, 2 Superior Maxillary, 1 Inferior Maxillary, 2 Lachrymal, 2 Nasal, 1 Vomer, 2 Turbinated, and 2 Palate.

152. *Enumerate the bones of the Ears.*

There are six bones in the ears, three in each : Maleus, Incus, and Stapes.

153. *How are the bones of the Trunk divided?*

The bones of the trunk are divided into Spinal Column, Ribs, Sternum, Os Hyoides, and Pelvis.

154. *What is the Spinal Column?*

The Spinal or Vertebral Column is the central axis of the body, and consists of twenty-four bones, or *Vertebræ*, excluding the Sacrum and Coccyx, which are sometimes reckoned as a part of it.

155. *Into what three classes are the Vertebræ divided?*

The twenty-four vertebræ are grouped as follows : seven Cervical, twelve Dorsal, and five Lumbar.

156. *What are the first two cervical vertebræ called?*

The first two cervical vertebræ have received special names, the first being called the *Atlas*, and the second, the *Axis.*

157. *Describe the mechanical structure of a vertebra.*

A vertebra consists of a short, thick cylindrical *Body*, from which branch backward two *Pedicels* uniting behind by two *Laminæ* to form a *Neural Arch.* From this Neural Arch, or ring, project three prominent processes for the attachment of muscles, two laterally and one backwards.

158. *What is the Sacrum?*

The Sacrum is a large, wedge-shaped bone, situated at the base of the spinal column, at the back part of the pelvic cavity between the *Innominata.* This bone, which is properly a continuation of the spinal column, consists in youth of five distinct vertebræ, which unite into a single bone later in life.

159. *What is the Coccyx?*

The Coccyx is a small, curved prolongation of the Sacrum ; and like that bone, it consists in early life of distinct vertebræ, four in number.

160. *Why is the spinal column curved?*

Its curvature secures elasticity combined with strength, and tends to lessen the jarring of the brain.

161. *What other arrangement is found in the spinal column for the protection of the brain?*

Pads of cartilage are placed between the separate vertebræ, thus deadening the shock of running, jumping, etc.

162. *What are Ribs?*

The Ribs are twenty-four long, slender, curved bones forming the walls of the thorax.

163. *Into what two classes are the ribs divided?*

True and False.

164. *What are the True Ribs?*

The first seven on each side are called True Ribs. Each is attached to the Sternum in front directly by its own separate cartilage.

165. *What are the False Ribs?*

The last five on each side are called False Ribs. The first three of each group are attached to each other at their anterior extremities by cartilage, which also joins them collectively to the Sternum ; the last two, having no anterior attachment, are called *Floating Ribs.*

166. *What is the Sternum?*

The Sternum is a flat bone situated in the median line in front of the thorax. It articulates with the collar bone and the ribs.

167. *Into what three parts is the Sternum divided?*

From its resemblance to an ancient sword, the three parts of the sternum have been called the *Manubrium*, or handle, the *Gladiolus*, or blade, and the *Ensiform Appendix*, or sword point.

168. *What is the Hyoid Bone?*

The Hyoid Bone, *Os Hyoides*, is a u-shaped bone situated at the base of the tongue, and having no articulation with the other bones of the skeleton.

169. *What are the Innominata?*

The Innominata are two large, irregular bones which with the Sacrum form the pelvic basin.

170. *Into what three parts is each Innominatum divisible?*

The Ilium, the Ischium, and the Pubes. These three divisions are separate bones in youth, becoming united into a single bone in middle life.

171. *What is the cup-like depression in the outer side of the Innominatum which receives the head of the Femur called?*

The Acetabulum, or vinegar cup. It is also called by anatomists the *Cotyloid Cavity.*

172. *What are the four divisions of the upper extremity?*

Shoulder, Arm, Forearm, and Hand.

173. *What two bones form the frame-work of the Shoulder?*

The Scapula, or "Shoulder Blade," and the Clavicle or "Collar Bone."

174. *Describe the Scapula.*

The Scapula is a large, flat, triangular bone, situated in the upper and back part of the thorax. It forms the principal support for the arm.

175. *What two well marked processes project from the outer angle of the Scapula?*

The Acromion and the Coracoid. The Acromion Process is the highest point of the Shoulder.

176. *What is the Glenoid Cavity?*

The Glenoid Cavity is the shallow socket into which the head of the Humerus is articulated.

177. *Describe the Clavicle.*

The Clavicle is a long bone, shaped like the stem of an Italic small *f.* It extends from the upper part of the Sternum outward and backward to the Acromion Process of the Scapula.

178. *What is the Humerus?*

The Humerus is the bone of the arm, extending from the shoulder to the elbow.

179. *What are the two bones in the Forearm called?*

Radius and Ulna. The radius is on the side next the thumb.

180. *What is the Olecranon Process?*

The Olecranon Process is a large eminence situated at the upper extremity of the Ulna. It can be readily perceived at the back of the elbow.

181. *What is the "Crazy Bone"?*

The "Crazy Bone," or "Funny Bone," is *not* a bone, but a nerve, which when compressed against the bone at the elbow produces a prickling sensation in the third and fourth fingers.

182. *Into what three groups are the bones of the Hand divided?*

Carpus, or bones of the wrist; Metacarpus, or bones of the hand proper; and Phalanges, or bones of the fingers.

183. *Enumerate the bones of the Carpus.*

There are eight bones of the Carpus, arranged in two rows, as follows first row—Scaphoid, Semilunar, Cuneiform, and Pisiform: second row—Trapezium, Trapezoid, Os Magnum, and Unciform.

184. *How many bones are there in the Metacarpus?*

Five, arranged in a single row in the palm of the hand.

185. *How many bones are there in the Phalanges?*

There are fourteen bones in the Phalanges, arranged in three rows, the thumb having but two bones.

186. *What are the three divisions of the lower extremity.*

Thigh, Leg, and Foot.

187. *What is the bone of the thigh?*

The Femur, the longest bone of the body.

188. *What are the Trochanters?*

The Trochanters are two prominent processes at the upper extremity of the Femur, designed for the attachment of the muscles that rotate the thigh.

189. *What are the two bones of the Leg?*

Tibia and Fibula. The Tibia, commonly called the " Shin Bone," is the larger and stronger, and is situated in the front and inner part of the leg.

190. *What is the Patella?*

The Patella, or " Knee Pan," is a large sessamoid bone lying in front of the knee joint, and designed to change the direction of the application of the force of certain muscles.

191. *What are the divisions of the bones of the Foot?*

Tarsus, Metatarsus, and Phalanges.

192. *Enumerate the bones of the Tarsus.*

The Tarsus, corresponding to the Carpus of the upper extremity, is composed of seven bones, as follows : Os Calcis, Astragalus, Cuboid, Scaphoid, Internal Cuneiform, Middle Cuneiform, and External Cuneiform.

193. *What is the Os Calcis?*

The Os Calcis is the large bone projecting backward to form the heel.

194. *How many bones are there in the Metatarsus?*

There are five bones in the Metatarsus, arranged in a single row similar to the Metacarpus, and forming the arch of the instep.

195. *How many bones are there in the Phalanges of the foot?*

Fourteen, arranged similarly to those in the hand.

196. *What are the functions of the skeleton ?*

The skeleton, as a frame, gives form to the body; it supports the various organs in their positions; it forms cavities for the protection of the vital organs; and its bones serve as levers in producing motion.

197. *What are the functions of the skull?*

The skull is the brain-box in whose thick, strong walls are carefully packed the organs of hearing, seeing, smelling, and tasting.

198. *What are the functions of the spinal column?*

The spinal column is the central axis for the support of the body, and has a strong, tubular cavity for the protection of the spinal cord.

199. *What are the functions of the ribs?*

The ribs, together with the spinal column and the sternum, form a strong, light cage for the protection of the heart and lungs, and, acting as levers, aid in respiration.

200. *What are the functions of the Pelvis?*

The Pelvis is a broad, strong ring swung between the upper extremities of the femurs. It supports the body through the spinal column, serves for the attachment of the strong muscles of the trunk and lower extremities, and provides a cavity for the protection of some of the viscera.

201. *What are the functions of the bones of the extremities?*

The bones of the extremities act as levers in producing the great variety of movements for which the limbs are designed.

202. *Contrast the structure of the upper and lower extremities, as determining their fitness for performing their special functions.*

The upper extremities are prehensile organs, and their mechanical arrangement is such as to permit great freedom of movement; the lower extremities support the weight of the body and are required to make fewer and simpler movements, and are, therefore, strong and compact. This contrast, noticeable throughout the limbs, is specially marked in the foot and hand. The fingers have much greater freedom of movement than the toes, the thumb admitting of being brought opposite each finger, a thing not possible with the great toe, even in the uncrippled feet of savages.

203. *What is the breaking of a bone called?*

A Fracture, which may be either a *Simple Fracture*, affecting only the bone, or a *Compound Fracture*, in which the surrounding integuments and muscles are lacerated.

204. *How is a broken bone reunited?*

If the broken parts are brought properly together and kept in that condition, new bone matter will be rapidly built into the breach until complete union takes place.

205. *Why are the bones of aged persons more easily broken than those of children?*

In later life the proportion of mineral matter in the bones largely increases, rendering them more brittle and more difficult to heal when broken.

206. *Why should not children in school be kept sitting on seats so high that their feet are not supported?*

The bones of young children are so flexible on account of their cartilaginous state that the weight of their dangling feet will bend the thigh bones over the edge of the seat and produce deformity.

207. *How are "Bow-legs" produced?*

Bow-legs are commonly produced by causing infants to support the weight of their bodies upon their limbs before their bones have hardened.

208. *What is "The Rickets"?*

The Rickets is a disease of the bones in which there is lack of mineral matter, and a consequent crookedness and distortion of the bodily frame. It is found in young children born of diseased parents in unhealthy neighborhoods.

209. *What is a Bone Felon?*

A Bone Felon, or Whitlow, is an inflammatory tumor occurring in the fingers near the joints, between the periosteum and the bone.

Anterior View of the Muscles of the Body.
1, Frontal Bellies of the Occipito-Frontalis.
2. Orbicularis Palpebrarum. 3, Levator Labii
Superioris Alæque Nasi. 4, Zygomaticus Minor. 5, Zygomaticus Major. 6, Masseter. 7,
Orbicularis Oris. 8, Depressor Labii Inferioris.
9, Platysma-Myoldes. 10, Deltoid. 11, Pectoralis Major. 12, Axillary portion of the Latissimus Dorsi. 13, Serratus
Major Anticus. 14, Biceps
Flexor Cubiti. 15, Anterior
Portion of the Triceps Extensor Cubiti. 16, Supinator
Radii Longus. 17, Pronator
Radii Teres. 18, Extensor

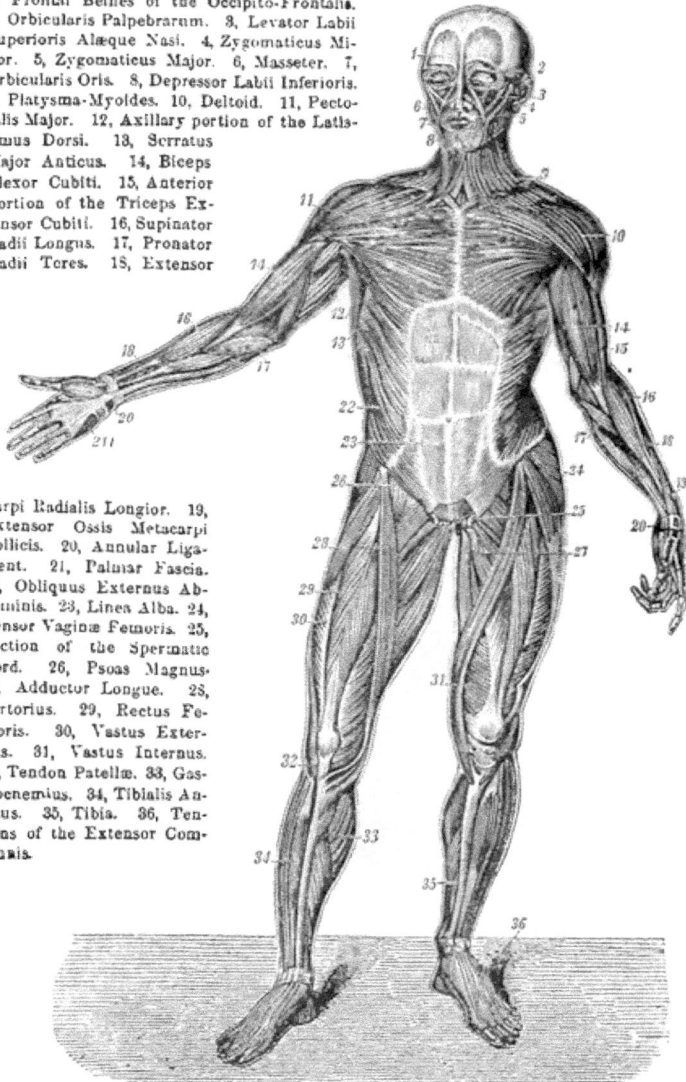

Carpi Radialis Longior. 19,
Extensor Ossis Metacarpi
Pollicis. 20, Annular Ligament. 21, Palmar Fascia.
22, Obliquus Externus Abdominis. 23, Linea Alba. 24,
Tensor Vaginæ Femoris. 25,
Section of the Spermatic
Cord. 26, Psoas Magnus.
27, Adductor Longue. 28,
Sartorius. 29, Rectus Femoris. 30, Vastus Externus. 31, Vastus Internus.
32, Tendon Patellæ. 33, Gastrocnemius. 34, Tibialis Anticus. 35, Tibia. 36, Tendons of the Extensor Communis.

MUSCULAR SYSTEM.

Didactic Note.—In teaching the mechanical structure of muscles, the teacher must again resort to the butcher's shop for his materials. A judicious use of the "living model" is of the highest importance. The Deltoid muscle becomes a real existence, instead of a mere name, to the boy or girl who has felt it on his or her own shoulder. Great discretion and tact, however, are necessary in such teaching. Topical outlines can be used to advantage in this division of the subject; but no effort must be made to teach a full outline of the muscles, such as is found in the larger Anatomies. A few important muscles thoroughly taught both as to location and functions should satisfy the most ambitious teacher. The grade of the class will determine the amount of time to be given to the study of contraction and muscular work. Topics in the hygiene of muscles will be found in abundance by the wide-awake teacher.

SPECIAL OUTLINE.

2⁴ Muscular System. [See General Outline, page 7.]

  1⁵ General study of muscles.

    1⁶ Structure.

      1⁷ Parts.

        1⁸ Belly, etc.

      2⁷ Coverings: Fascia.

      3⁷ Attachments.

        1⁸ Tendons, etc.

    2⁶ Classes.

      1⁷ As to form.

        1⁸ Radiate, etc.

      2⁷ As to position.

        1⁸ Superficial, etc.

      3⁷ As to direction of motion.

        1⁸ Abductors, etc.

    4⁷ As to control by the will.

      1⁸ Voluntary, etc. [tion.

    5⁷ As to concurrence of ac-

      1⁸ Congenerous, etc.

    3⁶ Action.

      1⁷ Muscular contraction.

      2⁷ Measure of work. [age.

      3⁷ Mechanical disadvant-

  2⁵ Analysis of the system.

    1⁶ Number of muscles.

    2⁶ Naming of muscles.

    3⁶ Grouping of muscles.

1⁷ Of the head.

1⁸ Occipito-frontalis, etc.

2⁷ Of the neck.   [etc.

1⁸ Sterno-cleido-mastoid,

3⁷ Of the trunk.

1⁸ Pectoralis major, etc.

4⁷ Of the upper extremity.

1⁸ Deltoid, etc.

5⁷ Of the lower extremity.

1⁸ Sartorius, etc.

3⁵ Functions.

1⁶ Maintenance of postures.

2⁶ Locomotion.

3⁶ Vital actions.

4⁶ Muscular Sense.

5⁶ Expression of thought and feeling.

4⁵ Hygiene.

210. *What is the Muscular System?*

The Muscular System is a collection of more than five hundred distinct contractile bodies, which in their proper action produce the great variety of movements belonging to the different parts of the body.

211. *What is a Muscle?*

A Muscle is a body of contractile tissue, constituting a separate motor organ with a distinct office.

212. *What are the parts of a muscle?*

Most muscles may be divided into the Belly, or thick, fleshy part, and the Extremities.

213. *What are Fasciæ?*

The Fasciæ are strong, inelastic membranes, investing the muscles and binding their fasciculi firmly together.

214. *How are muscles attached to the bones and integuments?*

From the central belly toward the extremities muscular fibres change gradually into fibres of white connective tissue constituting bands for attachment to the bones and integuments.

215. *What is the origin of a muscle?*

The origin of a muscle is the fixed central attachment toward which it pulls.

216. *What is the insertion of a muscle?*

The insertion of a muscle is the outer movable point on which its action is exerted.

217. *What determines the application of the terms origin and insertion as applied to the points of attachment of muscles?*

These terms as applied to the attachments of most muscles are purely relative, the origin in one action becoming the insertion in another. In the description of muscles it is customary to denote by the term "origin" that extremity which is *usually* the more fixed.

218. *What are Aponeuroses?*

Aponeuroses are broad, flattened bands of white fibrous connective tissue joining muscles to bone and integuments.

219. *What are Tendons?*

Tendons are slender, cord-like bands, similar in structure and use to aponeuroses.

220. *Into what classes are muscles divided with respect to form?*

Radiate, Peniform, Sphincter, Hollow, etc.

221. *Into what classes are muscles divided with respect to position?*

Superficial and Deep-seated.

222. *Into what classes are muscles divided with respect to direction of motion?*

Abductors and Adductors, Pronators and Supinators, Flexors and Extensors.

223. *Into what classes are muscles divided with respect to control by the will?*

Voluntary, Involuntary, and Mixed.

224. *Into what classes are muscles divided with respect to concurrence of action?*

Congenerous and Antagonistic.

225. *Describe the action of muscles.*

Muscles do their work by shortening themselves in the direction of the longitudinal extension of their fibres, thus drawing more nearly together the parts to which they are attached.

226. *What time is required for muscular contraction?*

The rapidity and precision of muscular contraction is one of the wonders of the human body. Many of the changes in the vocal organs in rapid speaking are accomplished in less than $\frac{1}{50}$ of a second; and in the vocalization of an accomplished singer there is a beautiful exactness in the movements that is equally marvelous.

227. *What is the measure of the work done by a muscle?*

In determining the amount of work done by a muscle both the weight moved and the distance through which it is moved are considered. If a muscle by its contraction raises four pounds through five feet against gravity it is said to do twenty "foot-pounds" of work.

228. *Upon what does muscular strength depend?*

Muscular strength depends upon a healthy state of the body in general and upon judicious exercise.

229. *What is Fatigue?*

Fatigue is a state of physical exhaustion, resulting from violent and protracted exercise, in which the various functions are performed with less vigor and promptness than usual.

230. *What is the immediate cause of fatigue?*

Fatigue results from the disintegration, or breaking down, of muscular tissue more rapidly than it can be repaired by the various constructive agents.

231. *What is meant by the "mechanical disadvantage" at which muscles act?*

On account of the oblique direction in which muscles commonly pull, much of their force is lost. This mechanical disadvantage, as it is called, results from the manner of insertion, and is very great in nearly all the muscles of the body.

232. *What is the design in arranging muscles so as to act at so great loss of available force?*

The mechanical disadvantage, which is such as regards mere power only, results from a sacrifice of strength to rapidity, and of directness in the application of force to compactness of structure. Compare the Masseter muscle of the jaw, where strength alone is required, with the Biceps of the arm, where rapidity in movement of the hand is desired: the Masseter pulls at right angles, with but little lost force; but the Biceps acts very obliquely upon the forearm as a lever of the third class, producing quickness of movement at a great loss of available force.

233. *By what plan is this mechanical disadvantage partly overcome?*

The mechanical disadvantage is in part overcome by the enlargement of the joints, which causes the tendons to pull over the projections, and, consequently, renders the action of the muscle more direct. The Patella at the knee-joint is a beautiful contrivance of this kind. Pulleys are also used to change the direction of the application of a force, as in the annular ligaments at the wrist and ankle, and in the Digastric muscle of the jaw.

234. *How many muscles are there in the human body?*

There are 527 distinct muscles in the body, arranged in 261 pairs and 5 single ones.

235. *What are atrophied muscles?*

Atrophied muscles are muscles wasted away from disease or disuse until they are no longer able to perform their functions. The three little muscles attached to the external ear, as if to move it like the ear of an animal, are rudimentary or atrophied by disuse.

236. *How are muscles named?*

Muscles are named from their *locations*, as Interspinales, etc.; from their *forms*, as Rhomboideus, etc.; from their *attachments*, as Sterno-cleido-mastoid, etc.; from their *use*, as Diaphragm, etc.; and from their *directions*, as Transversales, etc.

237. *How are muscles grouped in descriptive anatomy?*

Muscles are grouped for convenience of study in various ways; but the most convenient arrangement is with reference to location, as muscles of the *head, neck, trunk, upper extremities* and *lower extremities.*

238. *How is a muscle described?*

A muscle is commonly described by giving its location, form, attachments, and action. A shorter description simply tells where it is and what it does.

239. *Describe the Occipito-frontalis.*

The *Occipito-frontalis* is one of a pair of muscles that cover the entire crown from the back of the head to the eyebrows. It raises the eyebrows, wrinkles the forehead, and moves the scalp.

240. *Describe the Corrugator supercilii.*

The *Corrugator supercilii* is a small muscle at the inner end of the eyebrow. It draws the eyebrow inward and downward in frowning.

241. *Describe the Orbicularis palpebrarum.*

The *Orbicularis palpebrarum* is a sphincter muscle surrounding the eye. It closes the lid, either voluntarily or involuntarily.

242. *Describe the Orbicularis oris.*

The *Orbicularis oris* is an eliptical sphincter muscle surrounding the mouth. It closes and puckers the mouth.

243. *Describe the Levator labii superioris alæque nasi.*

The *Levator labii superioris alæque nasi* is a slender muscle extending from the inner margin of the orbit of the eye to the nostril and the upper lip. It lifts the upper lip and dilates the nostril in expression of disgust.

244. *Describe the Temporal muscle.*

The *Temporal* muscle is a broad, radiating muscle situated on the side of the head in front of and above the ear and attached below to the lower jaw. It is used in biting and chewing.

245. *Describe the Masseter.*

The *Masseter* is a short, thick muscle situated below the ear at the angle of the lower jaw. It is used in chewing.

246. **Describe the** *Buccinator*

The *Buccinator* is a broad, thin muscle situated in the cheek in front of the *Masseter.* It compresses the cheek to expel air forcibly in blowing.

247. *Describe the Zygomaticus major.*

The *Zygomaticus major* is a slender muscle extending from the Malar bone to the angle of the mouth. It is used in laughing.

248. *Describe the Sterno-cleido-mastoid.*

The *Sterno-cleido-mastoid* is a large, thick muscle extending obliquely upward and backward at the side of the neck, from the Sternum and Clavicle in front to the mastoid process of the Temporal bone behind the ear. It bends the head forward and rotates it toward the opposite side.

249. *Describe the Trapezius.*

The *Trapezius* is a broad, triangular muscle which with its fellow covers the upper and back part of the neck and shoulders. It draws the head backward, or elevates the point of the shoulder.

250. *Describe the Latissimus dorsi.*

The *Latissimus* **dorsi** is a large, flat muscle extending from the lumbar and lower **dorsal** vertebræ obliquely upward and forward to the Humerus. It draws the arm downward and backward, turning it inward.

251. *Describe the* **Erector spinæ.**

The *Erector spinæ* is a large, muscular mass lying along the spinal column. It keeps the body erect and bends it backward.

252. *Describe the Pectoralis major.*

The *Pectoralis major* is a broad, thick, triangular muscle extending from the inner end of the Clavicle and upper part of the Sternum to the Humerus. It draws the arm downward and forward across the chest.

253. **Describe** *the Serratus magnus.*

The *Serratus magnus* is a broad, thin muscle situated at the side of the chest. It draws the Scapula forward and inward.

254. **Describe the** *Intercostals.*

The *Intercostals* are thin layers of muscles filling the spaces between the ribs. The external layer extends obliquely downward and forward, and raises the ribs; the internal layer extends downward and backward, and depresses the ribs.

255. *Describe the Diaphragm.*

The *Diaphragm* is a thin, muscular wall separating the thorax from the abdomen. It is used in breathing.

256. *Describe the Deltoid.*

The *Deltoid* is a large, triangular muscle situated on the upper and outer part of the arm and shoulder. It raises the arm from the side.

257. *Describe the Biceps of the arm.*

The *Biceps* of the arm, or the *Biceps flexor cubiti,* is a long, thick, spindle-shaped muscle situated in the front of the arm. It arises by two heads from the Scapula and extends downward to the Radius. It bends the forearm upon the arm.

258. *Describe the Triceps.*

The *Triceps* is a large muscle situated on the back of the arm. It arises by three heads from the Scapula and upper portion of the Humerus and extends downward to the Olecranon process of the Ulna. It extends and straightens the forearm upon the arm.

259. *Describe the Pronator radii terres.*

The *Pronator* is a small muscle extending obliquely across the arm below the elbow. It turns the Palm of the hand down.

260. *Describe the Supinator longus.*

The *Supinator* is a long muscle extending from the lower end of the Humerus along the Radial side of the arm to the Radius at the wrist. It turns the palm of the hand up.

261. *What are the muscles which are used in closing and opening the hand.*

The muscles used in closing and opening the hand are called *Flexor* and *Extensor digitorum.*

262. *Describe the Gluteus maximus.*

The *Gluteus maximus* is a thick mass of coarse, muscular tissue extending from the Ilium obliquely downward and outward to the outer side of the Femur. It aids in keeping the body erect and rotates the thigh outward.

263. *Describe the Sartorius.*

The *Sartorius,* or "Tailor's Muscle," is the longest muscle of the body, extending obliquely downward and inward across the front of the thigh from the Ilium to the Tibia. It bends the leg upon the thigh and the thigh upon the body, rotating the thigh outward.

264. *Describe the Adductor muscles.*

The *Adductors* are three large muscles extending from the pubic region of the pelvis to the inner side of the Femur. They draw the thigh inward with great force.

265. *Describe the Rectus femoris.*

The *Rectus femoris* is situated along the front of the thigh, extending from the Pelvis to the Patella. It aids in supporting the body.

266. *Describe the Gastrocnemius.*

The *Gastrocnemius* is a large muscle constituting the greater part of the calf of the leg, and extending from the lower end of the Femur to the heel. It is used in walking.

267. *Describe the Soleus.*

The *Soleus* is a broad, flat muscle situated beneath the Gastrocnemius and extending from the upper extremities of the Tibia and Fibula to the heel. It is used in walking, and when the body is erect prevents it from falling forward.

268. *Describe the Tendo Achilles.*

The *Tendo Achilles* is the common tendon of the Gastrocnemius and Soleus forming their attachment to the *Os Calcis*. It is about six inches in length, and is the thickest and strongest tendon of the body.

269. *What muscles are brought into use in standing erect?*

It requires the action of many muscles to maintain the body in an erect position, chiefly the extensor muscles of the lower limbs and those that support the head and spine.

270. *What muscles are used in walking?*

Walking is accomplished by a succession of complicated movements of the flexor and extensor muscles of the legs.*

271. *How do we leap?*

In leaping the whole body is raised from the ground by the sudden and forcible extension of the lower limbs. In a long leap a preparatory crouching attitude is assumed by flexing the limbs.

272. *How is running accomplished?*

Running is a combination of rapid walking and leaping. Quick steps are taken, and at the same time the body is thrown forward by a succession of low, one-legged leaps. In running there is an instant between the leaps when both feet are off the ground at the same time.

273. *How are the "vital actions" maintained?*

Vital actions are the work of the involuntary muscles of the heart, walls of the intestines, etc. The processes of circulation, digestion and respiration are maintained by these muscles during the whole period of life without control by the will.

274. *What is the "Muscular Sense"?*

The term "Muscular Sense" denotes the faculty or capacity of determining by pressure and resistance the weight, texture, etc., of substances and the equipoise of the body. Grace of movement in dancing and skating are the result of the cultivation of this faculty.

275. *How are thought and feeling expressed?*

The thoughts and emotions of the mind are expressed entirely by muscular contraction. The muscles of the respiratory and vocal organs produce voice with its various intonations and inflections; the muscles of the hand produce the visible language of writing and drawing; and the muscles of the face express all shades of thought and feeling.

276. *How is Hiccough produced?*

Hiccough is produced by a sudden contraction of the diaphragm, which causes a rapid and forcible inhalation through the partially closed opening of the larynx.

277. *What are Convulsions?*

Convulsions are violent, involuntary contractions of those muscles that are usually controlled by the will. When the contractions are less violent, chronic, and marked by twitching of the muscles of the face and extremities, the disease is called "St. Vitus's Dance."

278. *What is "Locked-Jaw"?*

"Locked-Jaw" is a permanent contraction of the muscles, producing rigidity of the body and stiffening of the jaw.

279. *What is Rheumatism?*

Rheumatism is a shifting inflammatory affection of the muscles and joints. It is characterized by acute or burning pain with swelling of the affected part.

280. *What is Paralysis?*

Paralysis is a cessation of muscular activity resulting from a diseased state of the nervous apparatus. It may be either temporary or permanent; and, if permanent, it is accompanied by a wasting away of the affected muscles through disuse.

ARTICULATORY SYSTEM.

Didactic Note—The general structure and principal classes of joints are easily taught by observing the suggestions in the preceding notes. With younger children the teacher will do well to drop the scientific terminology, thus "Hinge joint" may be substituted for Ginglymus, etc. Special attention should be given to noting the adaptation of the various joints to their peculiar locations and functions.

SPECIAL OUTLINE.

3¹ Articulatory System. [See General Outline, page 8.]
  1² Structure.
    1³ Of Immovable Joints.
    2³ Of Movable Joints.
  2² Classification.
    1³ Synarthrosis.
      1⁴ Sutura.
      2⁴ Schindylesis.
      3⁴ Gomphosis.
    2³ Diarthrosis.
      1⁴ Arthrodia.

      2⁴ Enarthrosis.
      3⁴ Ginglymus.
    3³ Amphiarthrosis : Symphysis.
    3² Movements.
      1³ Gliding movement.
      2³ Angular movement.
      3³ Circumduction.
      4³ Rotation.
  4² Hygiene.

281. *What is an Articulation?*

Any union of two or more bones is called an articulation. The term "joint" is by some writers regarded as synonymous with "articulation," and by others as denoting a movable articulation only.

282. *Describe the structure of an immovable articulation.*

In immovable articulations the bones are brought into close contact, being separated only by a thin, fibrous membrane.

283. *Describe the general structure of movable articulations.*

In movable articulations the bones are slightly separated, their enlarged ends being covered by cartilage and joined together by strong ligaments. Each articulation is surrounded by a short cylindrical sack which secretes a lubricating fluid.

284. *What is Articular Cartilage?*

Articular Cartilage is that which covers the opposed ends of bones in articulations.

285. *What are Ligaments?*

Ligaments are strong bands of fibrous tissue which bind together the ends of bones in articulations.

286. *What is Synovial Membrane?*

Synovial Membrane is a delicate membrane surrounding the articulations and secreting the lubricating fluid.

287. *What are Synovial Bursæ?*

Synovial Bursæ are small membranous sacs situated at the articulations and secreting an oily fluid.

288. *What is Synovia?*

Synovia, or Synovial Fluid, is a transparent yellowish fluid like the white of an egg, slightly salt to the taste.

289. *Into what three classes are articulations divided?*

Synarthrosis, or immovable; *Diarthrosis*, or movable; and *Amphiarthrosis*, or mixed.

290. *What are Synarthrosis articulations?*

Synarthrosis articulations are those in which the bones are so closely joined as not to admit of any appreciable motion.

291. *What are the three varieties of Synarthrosis articulations?*

Sutura, Schindylesis, and Gomphosis.

292. *What are Sutura?*

Sutura are seams between closely joined bones, in which the edges are roughened and interlocked. When the edges are dovetailed together, the articulation is called a *true suture;* but when the edges are merely roughened, it is called a *false suture.* Various forms of sutures are found in the articulations of the bones of the skull and face.

293. *What are Schindylesis articulations?*

Schindylesis articulations are formed by inserting the thin edge of one bone in a groove in the edge of another, as in the articulation of the Sphenoid and Ethmoid.

294. *What are Gomphosis articulations?*

Gomphosis articulations are formed by inserting a conical point into a socket, as the articulation of the teeth with the jaw.

295. *What are Diarthrosis articulations?*

Diarthrosis articulations are joints admitting of considerable movement.

296. *What are the three varieties of Diarthrosis articulations?*

Arthrodia, Enarthrosis, and Ginglymus joints. **Other varieties are** given by anatomists; but these **are the** most important.

297. *What are Arthrodia?*

Arthrodia are joints formed by the opposition of plane surfaces and admitting a gliding movement, as in the wrist and ankle.

298. *What are Enarthrosis joints?*

Enarthrosis, or Ball-and-socket joints are formed by the reception of a globular head into a cup-like cavity, as in the hip and shoulder.

299. *What are Ginglymus joints?*

Ginglymus, or Hinge joints are those that admit of motion in only two directions, as in the elbow and knee.

300. *What are Amphiarthrosis articulations?*

Amphiarthrosis articulations are those in which the bones are joined by cartilage, and which admit of slight movement in all directions, as between the vertebræ.

301. *What four kinds of movements have joints?*

Gliding movement, Angular movement, Circumduction, and Rotation.

302. *Describe the Gliding movement.*

In the Gliding movement the surface of one bone slides over the surface of another without any angular or rotatory movement, as in the carpal and tarsal bones.

303. *Describe the Angular movement.*

The Angular movement is produced by the flexion and extension of one bone upon another, as in the hinge joints.

304. *Describe Circumduction.*

Circumduction is the movement of a long bone about one of its extremities as the apex of a cone, as in the shoulder and hip joints.

305. *Describe Rotation.*

Rotation is the turning of a bone upon its own axis, as the Atlas around the Odontoid process of the Axis.

306. *Describe the "long joint" made by the bones of the forearm.*

The hand is joined principally to the Radius, while the arm is joined chiefly to the Ulna; therefore, by the articulation of these two bones at their extremities, a joint is formed which may be considered to extend from the elbow to the wrist.

307. *What is a Dislocation ?*

A Dislocation is the displacement of a bone at a joint so as to partly or wholly destroy the articulation. Dislocated bones are said to be " out of joint."

308. *What is a Sprain?*

A Sprain is an injury of the ligaments of a joint, produced by violent twisting or straining. Sprains are very painful and heal slowly.

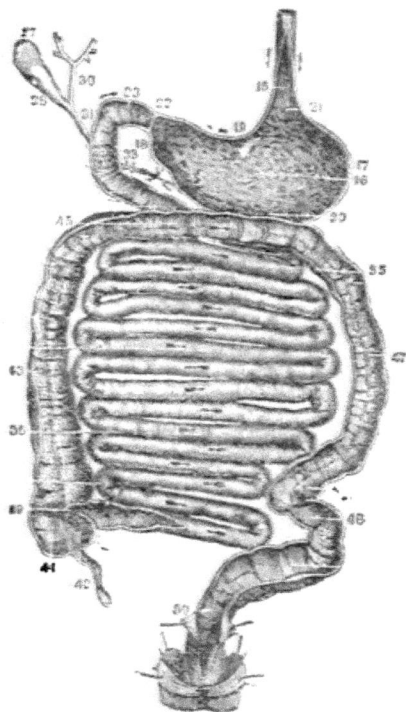

THE ALIMENTARY CANAL.

15, Œsophagus. 16. Stomach. 17, Splenic end of stomach. 18. Pyloric end of stomach. 19. Lesser curvature of stomach. 20, Greater curvature of stomach. 21, Cardiac orifice. 22, Pyloric orifice. 23, Duodenum. 27, Gall-bladder. 28. Cystic duct. 30, Hepatic duct. 31, Ductus communis choledochus. 33, Pancreatic duct. 35. Jejunum. 36. Ileum. 39. Ileo-cæcal valve. 41, Cæcum. 42, Vermiform appendix. 43. Ascending colon. 45. Transverse colon. 47, Descending Colon. 48. Sigmoid flexure. 50, Rectum.

DIGESTIVE SYSTEM.

Didactic Note.—Teach by outline and by the use of black-board drawings, the division, location, and form of the digestive apparatus. Fix in the mind by repeated drills the successive steps in the digestive process. The classification of foods should be brought within the comprehension of the pupils. Simplicity should not be sacrificed to logical exactness. Let the lessons on eating and drinking be as practical as they can be made, and repeated from time to time until they are thoroughly understood and appreciated. The teacher should carefully observe that no pet theories of his own are made unduly prominent.

SPECIAL OUTLINE.

1^4 Digestive System. [See General Outline, page 8.]

1^5 Apparatus.
1^6 Alimentary Canal.
1^7 General description.
2^7 Divisions.
1^8 Mouth.
2^8 Pharynx.
3^8 Œsophagus.
4^8 Stomach.
5^8 Intestines.
1^9 Small.
2^9 Large.
2^6 Accessory Glands.
1^7 Salivary Glands.
2^7 Liver.
3^7 Pancreas.
3^6 Supporting Membrane.
2^5 Foods.
1^6 General discussion.

1^7 Essential characteristics.
2^7 Variety necessary.
3^7 Alimentary principles.
4^7 Cooking.
5^7 Time of digestion.
6^7 Cured foods.
2^6 Classification.
1^7 Foods proper.
1^8 Organic.
1^9 Animal.
1^{10} Nitrogenous.
2^{10} Non-nitrogenous.
2^9 Vegetable.
1^{10} Nitrogenous.
2^{10} Non-nitrogenous.
2^8 Inorganic.
1^9 Solids.
2^9 Liquids.

2⁵ Auxiliary Foods.
  1⁵ Stimulants.
  2⁵ Condiments.
    1⁵ Saline.
    2⁵ Acid.
    5⁵ Pungent.

3⁵ Processes.
  1⁵ Mastication and Insaliva-
  2⁵ Deglutition.   [tion.
  5⁵ Chymification.
  4⁵ Chylification.
4⁵ Hygiene.

309. What is the Nutritive Apparatus?

The Nutritive Apparatus is a collection of systems which have for their joint function the maintenance and development of the body.

310. What five systems compose the Nutritive Apparatus?

Digestive, Absorptive, Circulatory, Respiratory, and **Excretory**.

311. What is the Digestive System?

The Digestive System is a collection of organs which prepare the food for the nourishment of the body.

312. What are the two great divisions of the apparatus of the Digestive System?

Alimentary Canal and Accessory Glands.

313. Describe the Alimentary Canal.

The Alimentary Canal is a musculo-membranous tube, about thirty feet in length, extending in a series of coils and convolutions from the mouth to the rectum.

314. What is the structure of the Alimentary Canal?

The Alimentary Canal consists of four coats, or layers: serous, muscular, areolar, and mucous. The serous coat is wanting in the upper portions.

315. What are the six divisions of the Alimentary Canal?

Mouth, Pharynx, Œsophagus, Stomach, Small Intestine, and Large Intestine.

316. Describe the Mouth?

The Mouth, or Buccal Cavity, is an irregular, oval-shaped cavity at the commencement of the Alimentary Canal. It is bounded in front by the Lips, on the sides by the Cheeks, above by the Hard Palate, below by a membranous floor extending from the Inferior Maxillary to the under side of the tongue, and behind by Soft Palate and Fauces.

317. *What is the capacity of the mouth?*

The mouth when fully open will contain about a half pint; when closed, the cavity is entirely filled by the tongue.

318. *What is the Palate?*

The Palate is the roof of the mouth. It consists of two parts: the Hard Palate in front, and the Soft Palate behind.

319. *What is the Uvula?*

The Uvula is a pendent, conical process of the soft palate, plainly visible in the back part of the mouth.

320. *What are the Tonsils?*

The Tonsils are two small glandular bodies, situated in the back part of the mouth on either side of the opening into the Pharynx.

321. *What are the Fauces?*

The term Fauces, or more properly the "Isthmus of the Fauces," denotes the opening between the mouth and the pharynx.

322. *What is the Pharynx?*

The Pharynx is that portion of the Alimentary Canal which is situated immediately behind the mouth and nose. It is about four and a half inches long and from one to two inches in diameter.

323. *What seven openings are in the walls of the Pharynx?*

Two to the Nostrils, *two* to the Eustachian Tubes, *one* to the Mouth, *one* to the Larynx, and *one* to the Œsophagus

324. *What is the Œsophagus?*

The Œsophagus is a tube about nine inches in length, extending from the pharynx to the stomach.

325. *Describe the structure of the Œsophagus?*

The Œsophagus consists of three coats: an outer *muscular*, middle *areolar*, and inner *mucous*. The muscular coat is composed of two layers of fibres; the outer arranged longitudinally and the inner transversely.

326. *Describe the Stomach.*

The Stomach is a large, pear-shaped expansion of the alimentary canal, situated on the left side, back of the lower ribs and beneath the diaphragm. It is about twelve inches in length, extending across the median line, four inches in width, and has a capacity of about a quart.

327. Of what are the walls of the stomach composed?

The walls of the stomach are composed of four coats: serous, muscular, areolar, and mucous.

328. Describe the muscular coat of the stomach.

The muscular coat of the stomach is composed of three layers of fibres. Those of the external layer extend *longitudinally* from the cardiac to the pyloric orifice; those of the middle, *transversely* around the stomach; and those of the inner, *obliquely* around the cardiac end.

329. Describe the mucous coat of the stomach.

The mucous coat of the stomach is an inelastic pink membrane having a peculiar honey-combed appearance produced by minute, shallow pits. It is large enough to admit of great distention of the stomach, and is, therefore, folded and wrinkled when the stomach is empty.

330. What are Gastric Follicles?

Gastric Follicles are minute glands opening into the bottoms of the shallow pits in the mucous coat of the stomach. One variety of these follicles is called *Peptic Glands.*

331. What is Gastric Juice?

Gastric Juice is a thin, colorless liquid, secreted by the gastric follicles.

332. What are the openings into the stomach?

The *Cardiac Orifice*, through which the food enters from the Œsophagus; and the *Pyloric Orifice*, through which it passes out to the small intestine.

333. What is the Pylorus?

The Pylorus, or Pyloric Valve, is a sphincter muscle formed of fibres of the middle layer of the muscular coat of the stomach. By its contraction the mucous coat is thrown into projecting folds, which prevent the exit of food during gastric digestion.

334. What is the Small Intestine?

The Small Intestine is a tortuous tube about twenty feet in length, extending from the stomach to the Large Intestine. It is divided arbitrarily by anatomists into three parts: the first, ten inches in length, is called the Duodenum; the second, about seven feet in length, is called the Jejunem; and the third, about twelve feet in length, is called the Ileum.

335. Describe the structure of the Small Intestine.

The Small Intestine consists of four coats, similar to those of the stomach, except that the muscular coat has but two layers of fibres and the mucous coat has a somewhat different structure.

336. *Describe the structure of the mucous coat of the small intestine.*

The mucous coat of the small intestine is a soft, pink, highly vascular structure, arranged in permanent transverse folds, and covered with minute processes called *Villi*, closely packed together like " pile " on velvet. In addition to the villi, numerous small glands open upon its surface.

337. *What are the functions of the villi?*

The villi project the mouths of the lacteals and blood vessels into the intestinal tube.

338. *What important glands discharge their secretions into the Duodenum?*

The Liver and the Pancreas.

339. *What is the Large Intestine?*

The Large Intestine is the last division of the alimentary canal. It is about five feet long and from one and a half to two and a half inches in diameter.

340. *What are the divisions of the large intestine?*

The large intestine is divided into Cæcum, Colon, and Rectum.

341. *Describe the Cæcum.*

The Cæcum is a large pouch at the commencement of the large intestine. It is really a projection of the Colon beyond the entrance of the Ileum.

342. *What is the Ileo-Cæcal Valve?*

The Ileo-Cæcal Valve is formed by two semi-lunar folds of mucous membrane at the entrance into the large intestine. It allows free passage toward the Cæcum, but prevents regurgitation.

343. *What is the Vermiform Appendix?*

The Vermiform Appendix is a small, coiled tube, about four inches long and as thick as a common lead pencil, joined to the Cæcum. It is thought to be a rudimentary prolongation of the Cæcum; but its use is not well understood.

344. *Describe the Colon.*

The Colon is a portion of the large intestine, about five feet in length, extending from the Cæcum to the Rectum. It presents for external examination a knotted, puckered appearance, caused by bands of muscular fibre shorter than the rest of the tube.

345. *What are the four divisions of the Colon?*

The *Ascending Colon*, extending from the Cæcum up along the right side of the abdominal cavity; the *Transverse Colon*, crossing from right to left below the stomach; the *Descending*

Colon, extending downward along the left side of the cavity; and the *Sigmoid Flexure*, an s-shaped curve where the Colon turns back upon itself and passes to the median line to join the Rectum.

346. *Describe the structure of the coats of the large intestine.*

The coats of the large intestine are the same as those of the small. The muscular coat has its external layer of fibres arranged in three longitudinal bands so attached as to pucker the tube; and the mucous coat has no *villi*, but is covered by numerous minute glands.

347. *What is the Rectum?*

The Rectum is the last division of the large intestine. It is about eight inches in length, and is nearly straight.

348. *What are Glands?*

Glands are organs whose functions are secretion, elaboration, and excretion. The ideal gland is a cellular mass separated into small lobes and pierced by minute tubes which unite into larger ducts through which the elaborated material is carried to the place of deposit.

349. *What are the Accessory Glands of the digestive apparatus?*

The principal glands of the digestive system are the Salivary Glands, the Liver, and the Pancreas. Besides these there are numerous minute simple bodies of similar function, as the Mucous Glands of the entire alimentary canal, the Gastric Follicles of the stomach, etc.

350. *What are the Salivary Glands?*

The Salivary Glands are minute glands imbedded in the walls of the buccal cavity. Their function is the secretion of saliva.

351. *What are the three pairs of salivary glands called?*

The three pairs of salivary glands are known as the *Parotid*, situated below and in front of the ears; the *Submaxillary*, in the floor of the mouth near the angles of the lower jaw; and the *Sublingual*, in the floor of the mouth farther forward than the Submaxillary.

352. *Through what passages do the secretions of the salivary glands reach the mouth?*

The secretion from the Parotid Glands flows to the mouth through two tubes known as *Steno's Ducts*, about two and a half inches long, discharging through easily-discerned openings in the cheeks opposite the second molar teeth of the upper jaw; that from the Submaxillary flows through two tubes about two inches in length, known as *Wharton's Ducts*, and discharge under the tip of the tongue; and that from the Sublingual flows through from eight to twenty small ducts from each gland, and discharges under the tip of the tongue.

353. *What is Saliva?*

Saliva is a thin, watery fluid secreted by the Salivary Glands. It is slightly viscid, and mingled with mucous as obtained in the spittle, is frothy.

354. *What is the Liver?*

The Liver is a large, reddish-brown gland, situated in the abdominal cavity immediately beneath the diaphragm, at the right of and overlying a portion of the stomach. It is the largest gland of the body, and weighs from three to four pounds.

355. *What are the two principal divisions of the liver?*

The liver is partially divided by a deep fissure into two bodies, called the Right and Left Lobes.

356. *Describe the structure of the liver?*

The liver has a soft, pliable texture, divided into minute masses, or *Lobules*, which are composed of *Hepatic Cells*.

357. *What are Bile Capillaries?*

Bile Capillaries are minute tubes extending among the hepatic cells, and having for their function the collecting of bile.

358. *What is the Gall Bladder?*

The Gall Bladder is a pear-shaped sac about four inches long and one inch in diameter, situated in immediate contact with the under surface of the liver. It is a reservoir for the bile when digestion is not in progress.

359. *What are the Biliary Ducts?*

The Biliary Ducts are, 1st, the *Cystic*, which conveys the bile from the bile capillaries to the gall bladder; 2nd, the *Hepatic*, which conveys the bile from the bile capillaries directly to the Common Bile Duct; and 3rd, the *Common Bile Duct*, which is formed by the junction of the Hepatic and Cystic ducts, and which discharges the bile into the Duodenum.

360. *What is the Bile?*

The Bile is a yellowish-green, viscid fluid secreted by the liver. It is extremely bitter and slightly alkaline to the taste.

361. *What is the Pancreas?*

The Pancreas is a long, tongue-shaped gland of yellowish color, situated behind the stomach.

362. *What is the Pancreatic Duct?*

The Pancreatic Duct is a small tube traversing the entire length of the Pancreas and conveying its secretion to the intestines. It usually has a mouth in common with the bile duct.

363. *Describe the Pancreatic Juice.*

The Pancreatic Juice is a viscid, alkaline fluid, slightly milky in appearance.

364. *What is the Peritoneum?*

The Peritoneum is a serous membrane lining the abdominal cavity, and, deflected from its wall, investing and supporting the viscera.

365. *What ligaments are formed by folds of the peritoneum?*

Those which support the liver, spleen, and other visceral organs.

366. *What is the Omentum?*

The Omentum, or " Caul," is a large, loose, double fold of the peritoneum, which protects and supports the viscera.

367. *What is the Mesentery?*

The Mesentery is a broad fold of the peritoneum, investing and supporting the small intestine.

368. *What are Foods?*

Foods are substances which, being digested in the body, furnish the proper materials for its development and the maintenance of its vital functions.

369. *What are the essential characteristics of foods?*

A substance in order to be classed as a food must either contain elements of which the body is composed, or fuel for maintaining its temperature; it must be digestible, *i. e.*, capable of being absorbed from the alimentary canal; and it must contain no substance injurious to the structure or activity of any organ.

370. *Why is variety necessary in food?*

Different kinds of food are necessary at each meal in order to furnish the proper elements for building up the various tissues of the body and to maintain the functional vigor and tone of the stomach.

371. *What are Alimentary Principles?*

Alimentary Principles are primary foodstuffs. They are not foods, but are simple substances of which foods are composed, and bear the same relation to foods that Proximate Principles do to Tissues.

372. *How are Alimentary Principles classified?*

Alimentary Principles, like Proximate Principles, are grouped into two classes, Inorganic and Organic. The Organic Alimentary Principles are divided into Nitrogenous and Non-nitrogenous.

373. *What are the most important Inorganic Alimentary Principles?*

Water, and the *chlorides, phosphates,* and *sulphates* of the bases, *sodium, potassium, magnesium* and *calcium.* These principles, with the exception of Water and the *chloride of sodium,* or common salt, are taken as food only in combination.

374. *What are the two classes of Nitrogenous Alimentary Principles?*

Proteids and Albuminoids.

375. *What are the sources of Proteid Principles?*

Proteids are obtained from both animals and plants. The most important are *Myosin* and *Syntonin* from lean meats, *egg albumen* from the white of eggs, *Casein* from milk and cheese, and *Gluten* from various plants.

376. *What are the sources of Albuminoids?*

Albuminoids are obtained mainly from the fibrous tissues of animals. The most important is *Gluten.*

377. *What are the two classes of Non-nitrogenous Principles?*

Hydrocarbons and Carbohydrates.

378. *What are the sources of Hydrocarbons?*

Hydrocarbons are obtained from animal fats and vegetable oils. The most important are *Stearin,* from tallow and suet, *Palmatin* from butter and lard, and *Olein,* from animal and vegetable oils.

379. *What are the sources of Carbohydrates?*

Carbohydrates are mainly of vegetable origin. The most important are *Starch, Sugar,* and *Gums.*

380. *What is the object of cooking foods?*

Cooking softens foods, breaks down their tissues, and renders them more easily masticated and digested. It also makes them more palatable by imparting to them agreeable flavors.

381. *What length of time is required to digest foods?*

From two to four hours, depending upon the kind of food, how it is prepared, the manner in which it is eaten, and the state of the system.

382. *What are cured foods?*

Cured foods are foods preserved by evaporating, salting, canning, etc. They are commonly not as palatable or nutritious as when fresh.

383. *What are the two great classes of Foods?*

Foods proper and Auxiliary Foods.

384. *What are the two classes of Foods proper?*

Organic and Inorganic.

385. *What are the two classes of Organic Foods?*

Animal and Vegetable.

386. *What are the two classes of Animal Organic Foods?*

Nitrogenous and Non-nitrogenous. The Nitrogenous group is composed of those foods in which nitrogenous alimentary principles predominate; the Non-nitrogenous of those in which the non-nitrogenous principles are in the excess.

387. *What are* the most *important Nitrogenous Animal Foods?*

Flesh, in all its varieties of bird, beast, and fish, *Eggs, Milk, Cheese,* etc.

388. *What are the most important Non-nitrogenous Animal Foods.*

Butter, Lard, Suet, etc.

389. *What are the two classes of Vegetable Organic Foods?*

Nitrogenous and Non-nitrogenous.

390. *What are the most important Nitrogenous* **Vegetable** *Foods?*

Cereal Grains, Beans, Peas, etc.

391. *What are the most important Non-nitrogenous Vegetable Foods?*

Root Plants, Fruit, Sugar Cane, etc.

392. *What are the most important Mineral Foods?*

Water and Salt.

393. *What are the uses of Water as a food?*

Water is a most important food, being used, 1st, to soften and dissolve other foods in the process of digestion; 2nd, to furnish material for the blood and other tissues of the body; 3rd, to hold substances in suspension while carrying them about the body, and washing away waste material; and 4th, to reduce the heat of the body by evaporation from its surface.

394. *What amount of Water is required daily as food?*

From three to five pints. This amount may may be taken entirely in combination in other foods.

395. *What are the uses of Common Salt as a food?*

Common Salt as a food proper is chiefly valuable as furnishing necessary elements for the Gastric Juice and Bile.

396. *What are Auxiliary Foods?*

Auxiliary Foods are alimentary substances containing but little nutritious matter, but which are chiefly valuable on account of their effects upon other foods, and upon the functional activity of the digestive and other organs.

397. *Into what two classes are auxiliary foods divided?*

Stimulants and Condiments. The dividing line is not sharply drawn, since condiments are, in a certain sense, stimulants.

398. *What are Stimulants?*

Stimulant Foods are those whose function is to increase vital activity or to strengthen and give tone to the system.

399. *What are the most important stimulant foods?*

Coffee, Tea, and Chocolate.

400. *How does Coffee act as a food?*

Coffee tranquillizes the nerves, clears the intellect, and relieves the sense of fatigue. The proper use of coffee rarely has deleterious effects ; but when such is the case, it should be abandoned.

401. *How does Tea act as a food?*

The action of Tea is similar to that of coffee. It also acts as a "negative food," delaying digestion.

402. *What is the value of Chocolate as a food?*

The effects of Chocolate are similar to those of coffee and tea, but in a less degree. It contains much more nutritious matter, and may be classed as a food proper.

403. *What are Condiments?*

Condiments are auxiliary foods whose function is to give relish to other foods. They contain but little nutritious matter.

404. *What are the three classes of Condiments?*

Saline Substances, Acids and Spices.

405. *What is the most important Saline Condiment?*

Common Salt, which, in addition to its use as a food proper, is a universal condiment. It stimulates taste and increases the flow of saliva.

406. *What is the most* **important** *Acid* **Condiment?**

Vinegar, an Acid fluid prepared **from many sources.** It flavors food and stimulates appetite.

407. *What are* *Spice* **Condiments?**

Spice **Condiments** are aromatic, pungent vegetable **substances** obtained from numerous sources, as fruit, flower, **root, leaf, bark,** and stem. The most important are Black Pepper, **Red Pepper,** Allspice, Nutmeg, Mustard, Clove, Cinnamon, Ginger. **Garlic. etc.**

408. *What is Digestion?*

Digestion is the **process of preparing food** in the alimentary canal for absorption **through its coats.** It consists in dissolving **the** food and **separating the nutritive** portions from the waste material.

409. *What* **are** *the steps in the digestive process?*

Mastication and Insalivation. Deglutition, Chymification **and** Chylification.

410. *In what part of the alimentary canal do Mastication* **and** *Insalivation take place?*

The operations of Mastication **and Insalivation are carried on** in the mouth.

411. *What is Mastication?*

Mastication, or "chewing, is the process of dividing **and** crushing the food by the teeth. The food is kept between the teeth by the joint action of the cheeks and tongue.

412. *What is the object of mastication?*

Food is broken by mastication **into** small particles so as to better enable the digestive fluids **of the** mouth and stomach to **act** upon it.

413. *What is Insalivation?*

Insalivation is the process of mingling **saliva and other** fluids of the mouth with food in mastication.

414. **What** *two kinds of functions has* **saliva?**

Mechanical and Chemical.

415. *What are the mechanical functions of saliva?*

It moistens food, making deglutition easier; dissolves it developing taste; and facilitates the mingling of the fluids o **the** stomach with it **when** it has passed to that organ.

416. *What are the chemical functions of saliva?*

It converts starch, an insoluble substance, into grape sugar, **a** soluble one. It also, when carried with the food into the stomach, excites the flow **of** gastric juice.

417. *What is Ptyalin?*

Ptyalin is that element of the saliva which turns starch into grape sugar.

418. *Describe Deglutition, or swallowing?*

After food has been properly masticated and mingled with saliva, it is forced to the back part of the mouth by the action of the cheeks and tongue, thence through the fauces and pharynx into the œsophagus by the rapid contraction of the muscles of the throat, and thence onward to the stomach.

419. *How do the muscles of the œsophagus act in deglutition?*

Food is forced through the œsophagus by the dilation and subsequent contraction of rings of muscular fibre, the fibres relaxing to permit its passage, and contracting above it to force it downward.

420. *In what part of the alimentary canal does Chymification, or Gastric Digestion, take place?*

In the Stomach.

421. *What is Chymification?*

Chymification is converting the masticated and insalivated food, received from the mouth, into Chyme in the stomach. It is both a mechanical and a chemical process.

422. *Describe the mechanical process of chymification.*

The stomach by the combined action of its three coats of muscular fibres maintains a sort of churning motion, called the "peristaltic movement," which converts the food into a semi-liquid mass thoroughly mixed with the gastric juice.

423. *Describe the chemical process of chymification.*

The gastric juice converts proteids and albuminoids into closely allied substances called Peptones, which are soluble and therefore capable of being absorbed through the coats of the stomach and intestines.

424. *What is Pepsin?*

Pepsin is that element in gastric juice which converts proteids into peptones.

425. *What is Chyme?*

Chyme is a grayish, semi-liquid substance, composed of peptones, gastric juice, saliva, starchy substances and indigestible portions of food.

426. *What is Chylification?*

Chylification is the process of converting chyme into chyle.

427. *Describe the process of chylification.*

After the chyme has passed through the pylorus into the duodenum, it is mingled with the bile and pancreatic juice, which convert it into a milky emulsion called chyle.

428. *What are Trypsin and Pancreatin?*

Trypsin and Pancreatin are active principles of the pancreatic juice. The former converts albuminoids into peptones; the latter converts starch into sugar.

429. *What is Chyle?*

Chyle is a milky emulsion composed of various quantities of all the elements of foods and products of digestion.

430. *What are Digestive Ferments?*

Digestive Ferments are the active principles of the various digestive fluids. They are, including those already named, Ptyalin, Pepsin, Trypsin, Pancreatin, Invertin, undetermined Bile ferments, and Curdling ferments of the gastric juice, pancreatic juice and intestinal fluids.

431. *What is Dyspepsia?*

Dyspepsia is a disordered state of the digestive system occurring in a variety of forms and degrees. It is not a well-defined disease; but is rather a class of symptoms that may indicate any one of several different diseased states.

432. *What are the causes of dyspepsia?*

Dyspepsia results from eating too much, eating too little, eating too fast, eating too rich food, eating too poor food, lack of bodily exercise, melancholy, etc.

433. *How should dyspepsia be treated?*

Where the disorder is great it should be treated only by a competent physician; but in most cases it is a mental rather than a physical disorder—*hypochondria*, in fact. It is in such cases a disease "pre-eminently suited for treatment by domestic quackery," and curable only by proper employment and the forgetting of self.

434. *What is "Biliousness"?*

"Biliousness," or a "Bilious Attack," is a disordered state supposed to result from either a deficiency or excess of bile. The term is applied ignorantly to several diseased states having nothing in common.

435. *What are "Mumps"?*

"Mumps" are a diseased state characterized by swelling of the parotid glands. They are quite painful; but if care is used to prevent taking cold, they soon run their course.

436. *Why is regularity of meals necessary for the maintenance of good health?*

The stomach works best when working by a program, with regular hours for labor and rest.

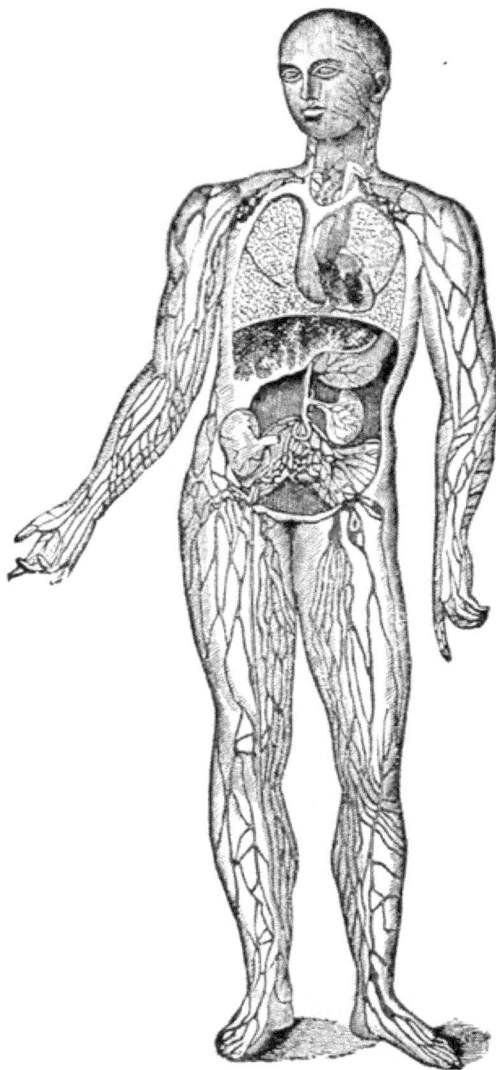

LYMPHATIC SYSTEM.

ABSORPTIVE SYSTEM.

Didactic Note.—Teach the general anatomy of the lymphatic apparatus, with special attention to the minute structure of the capillaries and lacteals. Good drawings, showing the microscopic structure of lacteals, are found in school physiologies, and should be copied on the blackboard. Develop a clear conception of the routes by which nutritious matter enters the circulation. Show by simple experiment the osmose of liquids, using a bladder filled with brine and immersed in a vessel of pure water. Impress the necessity of keeping the surface of the skin clean, and caution against poisoning.

SPECIAL OUTLINE.

2^4 Absorptive System. [See General Outline, page 8.]
 1^5 Apparatus.
  1^6 Blood Vessels [See Circulatory System].
  2^6 Lymphatic System.
   1^7 Thoracic Duct.
   2^7 Right Lymphatic Duct.
   3^7 Receptaculum Chyli.
   4^7 Lymphatics.
    1^8 General.
    2^8 Special: Lacteals.
   5^7 Lymphatic Glands.
    1^8 General.

  2^8 Special: Mesenteric
 2^5 Fluids.   [Glands.
  1^6 Lymph.
  2^6 Chyle.
 3^5 Processes.
  1^6 Absorption by blood capillaries.
  2^6 Action of Lymphatics.
  3^6 Action of Lacteals
  4^6 Osmose of Liquids.
  5^6 Propelling force.
 4^5 Hygiene.

437. *What is absorption?*
 Absorption is the process by which substances are taken up from the alimentary canal, and from the surface and other parts of the body, and conveyed into the circulation of the blood.

438. *What is the double mechanism of absorption?*

Absorption is the function of two distinct apparatuses, the veins and the lymphatic vessels. The structure of the veins is described under the circulatory apparatus.

439. *What is the apparatus of the absorptive system proper?*

The lymphatic vessels. Their sole function is absorption.

440. *Of what does the lymphatic apparatus consist?*

The Thoracic Duct, the Right Lymphatic Duct, the Receptaculum Chyli, the Lymphatic Vessels, the Lacteals, the Lymphatic Glands, and the Mesenteric Glands.

441. *Describe the Thoracic Duct.*

The Thoracic Duct is the main trunk of the lymphatic system. It is about the size of a common goose-quill, and from fifteen to eighteen inches long, extending from the Receptaculum Chyli opposite the second lumbar vertebra to the left subclavian vein.

442. *Describe the Right Lymphatic Duct.*

The Right Lymphatic Duct is a short tube which conveys the lymph from the right arm, right shoulder, and right side of the head and neck to the right subclavian vein.

443. *Describe the Receptaculum Chyli.*

The Receptaculum Chyli is an oblong reservoir into which lymph and chyle are collected, and from which they pass through the thoracic duct into the circulation of the blood.

444. *Describe the Lymphatic Vessels.*

The Lymphatic Vessels are slender, thread-like tubes, knotted at intervals by small glands. They absorb the lymph and carry it to the Receptaculum Chyli, Thoracic Duct, and Right Lymphatic Duct.

445. *What are Lacteals?*

Lacteals are the lymphatic vessels of the intestines. They carry chyle during the process of digestion.

446. *Describe the Lymphatic Glands.*

Lymphatic Glands are small, round, pinkish knots in the lymphatic vessels. They vary in size from that of a hemp-seed to that of an almond.

447. *Describe the passage of a lymphatic vessel through a lymphatic gland.*

As a lymphatic vessel approaches a lymphatic gland it divides into a number of small branches, called *Afferent Vessels*, which form a plexus in the substance of the gland, and after uniting, *within the gland*, into a single tube called an *Efferent Vessel*, pass out on the opposite side.

448. *What are Mesenteric Glands?*

Mesenteric Glands are the lymphatic glands of the lacteals.

449. *How are substances absorbed by the blood capillaries?*

A net-work of blood capillaries situated just beneath the surface of the mucous coat of the alimentary canal drinks in through thin walls, by the principle of " osmose of liquids," whatever is fitted for direct absorption into the blood.

450. *What substances are absorbed from the alimentary canal by the blood capillaries?*

Fluids and solids held in perfect solution, such as salts, sugars, etc., are taken directly into the circulation of the blood through the coats of the capillaries along the entire length of the alimentary canal.

451. *What is Lymph?*

Lymph is simply the plasma of the blood, containing a few white corpuscles and minute particles of the tissues in contact with which it is formed.

452. *How is Lymph formed?*

Blood, as such, being enclosed within continuous tubes, does not come into direct contact with the tissues of the body; but where it passes through the capillaries its plasma transudes through their thin walls and bathes the neighboring tissues. This transuded fluid, which contains a small number of white corpuscles, is called lymph.

453. *Describe the action of the Lymphatics.*

The Lymphatics originate in plexuses of minute tubes, which drink in through little mouths the transuded blood plasma, or lymph, and carry it back into the circulation.

454. *Describe the work done by the lymphatics in an inflamed spot.*

When a wound is made in any part of the body an unusual amount of blood is sent to the spot, bearing materials for the repair of the breach; and a correspondingly large amount of plasma and white corpuscles is forced through the strained walls of the capillaries, thus largely increasing the work of the lymphatics.

455. *What is Pus, or "Matter"?*

Pus is transuded plasma and white corpuscles gathered in an abcess at an inflamed spot. If it be healthy and be not meddled with by any lancing quackery, it will be rapidly carried back into the circulation by the lymphatics.

456. *How do the Lacteals act?*

Lacteals have little mouths in the villi of the small intestines, through which they absorb the chyle and carry it to the Receptaculum Chyli and Thoracic Duct. When digestion is not in progress, the lacteals carry lymph the same as other lymphatic vessels.

457. *What is the Osmose of liquids?*

Osmose, or Dialysis, is the process by which fluids of different composition and density, separated by a porous wall, pass through from side to side until the entire mass becomes homogeneous. It is by this process that the lymph gives materials to the cells of the tissues and takes materials from them.

458. *What is **the propelling force** of the lymphatic circulation?*

Lymph is forced onward through the lymphatic vessels toward the main trunks by pressure caused by the action of the respiratory apparatus in inspiration and expiration, the numerous valves in the lymphatic vessels allowing passage in one direction only. It is also crowded forward by new material constantly taken in by the little mouths of the vessels.

459. *Why is it necessary to keep the surface of the body clean?*

The skin has been called the " third lung " on account of its usefulness in exhalation and absorption; and good health depends upon its proper action, secured by keeping its pores open.

460. *How are persons poisoned by Wood-ivy?*

The poisonous exhalations of the ivy are absorbed through the skin. Contagious diseases are taken in the same way; so also medicines are administered by external application.

461. *What is Scrofula?*

Scrofula is a disease affecting the glands of the lymphatic system.

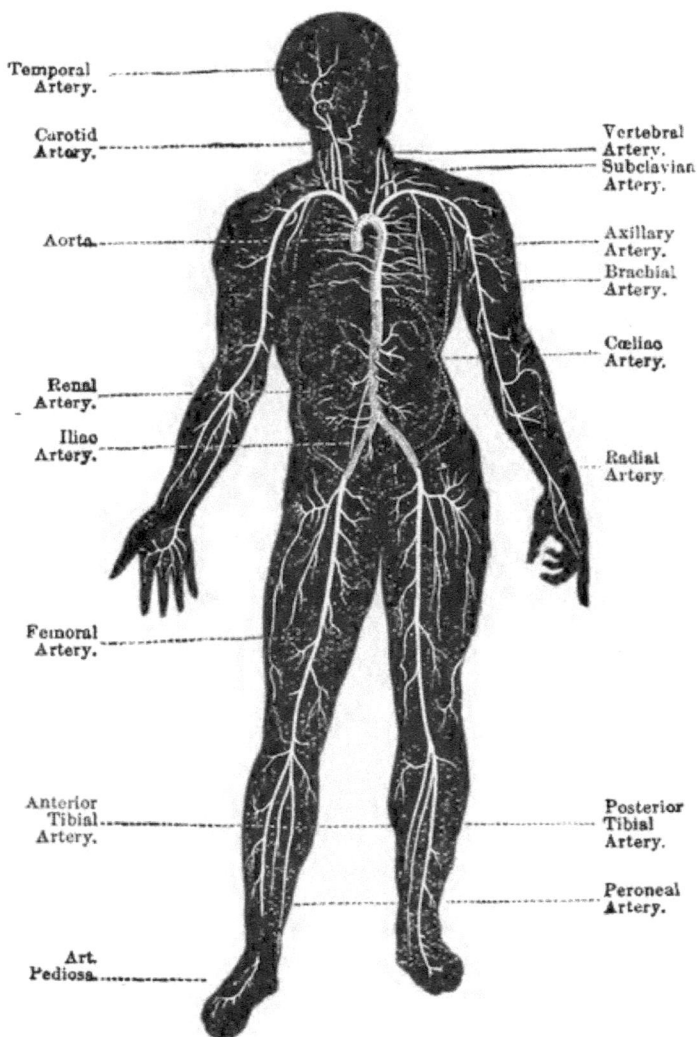

Temporal
Artery.

Carotid
Artery.

Aorta.

Renal
Artery.

Iliac
Artery.

Femoral
Artery.

Anterior
Tibial
Artery.

Art.
Pediosa.

Vertebral
Artery.
Subclavian
Artery.

Axillary
Artery.
Brachial
Artery.

Cœliac
Artery.

Radial
Artery.

Posterior
Tibial
Artery.

Peroneal
Artery.

Arterial System in Man.

CIRCULATORY APPARATUS.

Didactic Note.—This System is readily outlined, and with a proper use of the blackboard the general structure of its apparatus can be rapidly and thoroughly taught. Give special attention to the heart and capillaries. A beef's heart should be obtained from a butcher's shop and dissected in the presence of the class: its structure is very similar to that of the human heart. The use of a small compound microscope to show the corpuscles in the blood will add greatly to the interest of the study. The circulation of the blood through the capillaries in the web of the foot of a living frog can be shown with the same instrument. Drill the class individually and in concert, in tracing the circulation through the successive organs, until each pupil can follow it around from the left ventricle back to that cavity again. The functions of the blood and the changes that take place in it in the course of the circulation are important subjects of instruction.

SPECIAL OUTLINE.

3⁴ Circulatory System. [See General Outline, page 8.]

 1⁵ Apparatus.
  1⁶ Organs.
   1⁷ Heart.
    1⁸ General Description.
    2⁸ Parts.
     1⁹ Divisions.
      1¹⁰ Right.
      2¹⁰ Left.
     2⁹ Chambers.
      1¹⁰ Auricles.
      2¹⁰ Ventricles.
     3⁹ Orifices.
     4⁹ Valves.
      1¹⁰ Auriculo-Ventric-
      2¹⁰ Semi-lunar. [ular.
   2⁷ Arteries.
    1⁸ Structure.
    2⁹ Location.
   3⁷ Veins.
    1⁸ Structure.
    2⁸ Valves.
    3⁸ Location.
   4⁷ Capillaries.
  2⁶ Divisions.
   1⁷ Systemic.
    1⁸ Left Heart.

462. *What is the Circulatory System?*

The Circulatory System is a collection of organs whose function is the transportation of materials to the various parts of the body, as building material for the development and repair of the tissues, oxygen for chemical action, and ashes and waste matter for rejection from the body.

463. *How is transportation effected by the circulatory apparatus?*

The circulatory system transports materials by means of a carrying fluid forced through a system of tubes.

464. *What are the organs of the circulatory system?*

The Heart, the Arteries, the Veins and the Capillaries.

465. *Describe the Heart.*

The Heart is a hollow, conical, muscular organ, about the size of a man's fist, and weighing from ten to twelve ounces.

466. *Where is the heart situated?*

The heart is situated in the median line of the thorax, immediately back of the lower two-thirds of the sternum, and extends obliquely downward toward the left.

467. *Why is the heart commonly believed to be in the left side of the thorax?*

The apex of the heart lies on the left side opposite the space between the fifth and sixth ribs, and its striking against the walls of the thorax during the throbbing of the organ, causes the popular opinion as to the location of the entire heart in that region.

468. *Describe the structure of the heart?*

The heart is a hollow muscle, composed of several layers of anastomosing striated fibres. Its walls vary in thickness from one-twelfth to one-half inch, and enclose two entirely distinct cavities, each of which is separated by folding doors into two chambers.

469. *What is the Endocardium?*

The Endocardium is a delicate membrane which lines the cavities of the heart.

470. *What is the Pericardium?*

The Pericardium is a strong, double sack of fibro-serous membrane, which invests and supports the heart.

471. *What are the two principle divisions of the heart?*

The human heart is a double structure, consisting really of two distinct hearts, called the *Left* and the *Right Heart*, united into one organ.

472. *What are the four cavites of the heart?*

Each division of the heart has two cavites, or chambers, an *Auricle* and a *Ventricle*. The auricles occupy the upper portion of the heart; and the ventricles, the lower.

473. *What are the orifices of the heart?*

There are three sets of orifices in the heart: 1st, those through which the blood enters the heart, comprising two from the Venæ Cavæ into the right auricle and four from the Pulmonary Veins into the left auricle; 2nd, the two doorways between the auricles and ventricles; and 3rd, those through which the blood leaves the heart—one from the right ventricle to the Pulmonary Artery, and one from the left ventricle to the Aorta.

474. *What are the valves of the heart?*

The valves of the heart are little doors which permit free passage of the blood in one direction, but close to prevent regurgitation, or flow, in the opposite direction. Each division of the heart has two valves, one between the two chambers, and one where the blood leaves the ventricle.

475. *Describe the structure of the valves between the chambers of the heart.*

The valves between the chambers of the heart, called the *Auriculo-Ventricular Valves*, are flaps, or folds, of the endo-cardium, hanging from the margins of the *Auriculo-Ventricular*

Orifices into the ventricles. Their free edges are attached to the walls of the ventricles by cords—the *cordæ tendineæ*—long enough to permit their rising into contact with each other and closing the orifices.

476. *What are the Auriculo-Ventricular Valves called?*

The valve between the right auricle and right ventricle, having three flaps, is called the *Tricuspid Valve;* that between the left auricle and left ventricle, having two flaps, is called the *Bicuspid Valve.* The Bicuspid valve is also called the *Mitral Valve,* from a fancied resemblance to a Bishop's mitre.

477. *Describe the valves at the entrances of the Aorta and Pulmonary Artery.*

The entrances to the great arteries leading from the heart are guarded by valves called *Semi-lunar Valves,* formed by crescent-shaped folds, or pockets, which permit free exit of the blood from the ventricles and prevent regurgitation.

478. *What are Arteries?*

Arteries are strong, elastic tubes through which the blood passes from the ventricles to all parts of the body.

479. *Describe the general structure of arteries.*

The walls of all the larger arteries consist of three coats : a tough outer coat of white fibrous connective tissue, an elastic middle coat of involuntary muscular fibres and yellow elastic connective tissue, and a delicate inner coat of elastic membrane.

480. *Where are the arteries situated?*

The arteries are usually situated in protected cavities, and on the flexor side of the limbs, near the bones, and beneath the large muscles, that they may be subject as little as possible to pressure and wounding.

481. *What are Veins?*

Veins are tubes through which the blood returns to the heart after being forced out through the arteries.

482. *Describe the general structure of veins.*

Veins are similar in structure to arteries, except that the middle coat is less developed, rendering their walls thinner and more flaccid.

483. *Describe the valves of veins.*

Veins are provided with semi-lunar valves, formed by folds of the inner coat, and placed so that the mouths of the pockets are turned toward the heart. They permit free passage of the blood toward the heart, and prevent flow in the opposite direction.

484. *Where are the veins* **situated?**

Veins are usually situated in more exposed locations than the arteries. Deep-seated veins lie by the side of corresponding arteries, and **are**, therefore, called *vénæ comites*, or companion veins.

485. *What are Capillaries?*

Capillaries are microscopic tubes, about one-tenth of an inch in length, joining the ultimate divisions of the arteries to the commencement of minute veinlets. As the arteries divide again and again, becoming smaller and smaller, they finally pass into capillaries, which, uniting again into larger tubes, form the veins.

486. *Describe the structure of capillaries.*

As the arteries grow smaller the middle and **outer** coats gradually disappear, until in the capillaries only the inner coat remains, through whose delicate structure the plasma of the blood readily passes to the tissues.

487. *Into what two great divisions is the Circulatory System divided?*

Systemic, or **that of the** body in general, and Pulmonic, or that of the lungs.

488. *Which heart, or side of the heart,* **belongs to the systemic** *division of the circulatory apparatus?*

The left heart belongs to the systemic division **of the circu-**latory apparatus, and on account of the greater force required to drive the blood through the general system the walls of its ventricle are thicker than those of the ventricle of the right heart.

489. *What is the principal artery of the systemic division of the circulatory apparatus?*

The *Aorta*, a tube about an **inch** in diameter, **which arises** from the upper part of the left ventricle, extends **upward for** about two inches, arches backward toward the left **over the root** of the left lung, descends through the thorax along **the left side** of the vertebral column, pierces through the **diaphragm, and** extends in the abdomen to a point opposite **the fourth lumbar** vertebra, where it terminates by division into two **branches called** the *Right* and *Left Common Iliac.*

490. *What* **are the three parts or** *divisions of the aorta?*

Arch of the Aorta, Thoracic Aorta and Abdominal Aorta.

491. *What branches* **are** *given off from the arch of the aorta?*

Coronary, Innominate, Left Common Carotid, and Left Subclavian. The Innominate is very short, dividing into Right Subclavian and Right Common Carotid.

492. *What branches are given off from the Thoracic Aorta?*

No large branches are given off from the thoracic aorta; but many small ones run to the walls of the thorax, lung tissue, etc.

493. *What branches are given off from the Abdominal Aorta?*

Cœliac Axis, Superior Mesenteric, Renal, and Inferior Mesenteric. The Cœliac Axis, after extending about an inch, divides into three branches, the *Gastric, Hepatic,* and *Splenic,* supplying the stomach, liver, spleen, and pancreas; the two Mesenteric branches supply the intestines; and the Renal extend to the kidneys.

494. *Describe the arteries which carry the blood to the head.*

Blood is forced to the head through the *Right* and *Left Common Carotids* and the *Vertebral Arteries.* The Carotids pass upward on each side of the neck in front, dividing at the angle of the jaw into two branches, *External* and *Internal Carotid Arteries;* the external branches supply the face and scalp, and the internal extend to the brain. The *Vertebral Arteries* branch off from the *Subclavian Arteries,* and pass up through arches in the transverse processes of the six upper cervical vertebræ to the back part of the brain.

495. *What is the "Circle of Willis"?*

The *Circle of Willis* is a ring of arterial tubes formed at the base of the brain by anastomosing branches of the Carotid and Vertebral arteries. This arrangement furnishes two routes for the passage of blood to all parts of the brain, so that if one should be obstructed the other would be available.

496. *Describe the arteries that carry the blood to the upper extremities.*

Blood is forced to the upper extremities through the *Subclavian* arteries and their branches. A subclavian artery extends outward under the clavicle to the armpit, where it is called the *Axillary* artery, thence down along the inner side of the humerus, where it is called the *Brachial* artery; at the elbow it divides into two branches, called, from their positions, *Radial* and *Ulnar* arteries, which pass on to the hand, where they unite to form the *Palmar Arch?*

497. *Describe the arteries which carry blood to the lower extremities.*

The *Common Iliac* arteries, formed by the bifurcation of the Aorta opposite the fourth lumbar vertebra, divide, after extending about two inches, into the *External* and *Internal Iliac* arteries, the former carrying blood to the lower extremities, and the latter to the viscera and walls of the pelvis. The external division of the Common Iliac artery extends down along the front and inner side of the thigh to the knee, where it divides into two branches, the *Anterior* and *Posterior Tibial* arteries: the portion along the upper two-thirds of the thigh is called the *Femoral* artery; and that along the lower third and in the hollow of the knee joint, the *Popliteal.*

498. *What is the anastomosis of arteries?*

The *anastomosis* of arteries is their communication with each other by means of connecting branches, as in the Palmar Arch and the Circle of Willis.

499. *What are the two principal veins of the systemic division of the circulatory system?*

The *Superior* and *Inferior Venæ Cavæ*. The latter extends from opposite the fourth lumbar vertebra to the right auricle, and carries the blood from all parts of the body below the diaphragm to the heart; the former is a short trunk which carries blood from the whole upper half of the body to the right auricle.

500. *What is the Portal Vein?*

The Portal Vein is the main trunk of an auxiliary circulation: it carries the blood from the spleen, pancreas, and walls of the stomach and intestines to the liver, where it is distributed throughout the tissues of that organ.

501. *What heart, or side of the heart, belongs to the Pulmonary division of the circulatory system?*

The Right Heart belongs to the Pulmonary division of the circulatory system. It receives the blood from the body and pumps it into the lungs.

502. *What is the principal artery of the Pulmonary division of the circulatory system?*

The Pulmonary Artery, which extends upward for about two inches from the right ventricle and divides into two branches, called the Right and Left Pulmonary Arteries, which carry the blood directly to the lungs.

503. *What are the Pulmonary Veins?*

The Pulmonary Veins are four short trunks, which carry the blood from the lungs to the left auricle, entering that cavity by four separate orifices.

504. *Describe the arrangement of the pulmonary capillaries.*

The Pulmonary Artery, in its ultimate divisions, forms a dense net-work of capillaries on the walls of the air cells and air passages, bringing the blood into easy access to the purifying air currents.

505. *What is the Blood?*

The Blood is the nourishing fluid of the body ; it also constitutes a flowing current for the transportation of various materials.

506. *Give the physical characteristics of the blood?*

Blood from the arteries is scarlet, from the veins, dark purple ;
it is saline to the taste, clammy to the touch, and is slightly denser
than water.

507. *Describe the microscopic structure of the blood.*

The blood is composed of a nearly colorless, transparent fluid,
called the *Plasma*, in which float minute *Red* and *White Corpuscles.*

508. *What is the composition of the Plasma of the blood?*

The Plasma, or *Liquor Sanguinis*, is composed of a permanently fluid portion, the *Serum*, and certain albuminous substances, called *Fibrin Factors.*

509. *Describe the Red Blood Corpuscles.*

The Red Blood Corpuscles are circular biconcave discs, about
$\frac{1}{3200}$ of an inch in diameter and $\frac{1}{12000}$ of an inch in thickness.
It is only when collected in masses that they have a pronounced
red color; seen singly, each is a pale straw color.

510. *What gives the blood its color?*

The color of the blood is due to the presence of small particles of coloring matter, called *Hæmoglobin*, in the red corpuscles.

511. *What are White Blood Corpuscles?*

White Blood Corpuscles are minute, spherical bodies, rather
smaller than the red, and having all the characteristics of the
true animal cell.

512. *What changes are supposed to take place in the blood corpuscles?*

White blood corpuscles, which are identical with the corpuscles of the lymph and chyle, or but slight modifications of them,
are supposed to be gradually changed by some unknown process
into red corpuscles, which having served their purpose as oxygen carriers are thought to be destroyed in the spleen and liver..

513. *What are the two kinds of blood?*

Arterial, or bright red blood of the arteries, and *Venous*, or
dark purplish-red blood of the veins. The change in color is
due to the effect of oxygen upon the Hæmoglobin in the red
corpuscles.

514. *Describe Coagulation of blood.*

Blood drawn from the body separates in a short time into
two parts, a more or less solid *Clot*, and a fluid called *Serum.*

The Serum is a pale yellowish fluid, composed of water, albumen, and certain salts; the Clot is composed of interlacing threads of fibrin, containing in their meshes red and white corpuscles.

315. *What are the functions of the blood?*

The blood nourishes the tissues, conveys oxygen to places where it is needed for chemical action, maintains by its circulation uniformity of temperature throughout the body, supplies the various glands with the materials of their secretions, and carries the refuse and waste matters to the excretory organs for elimination from the body.

316. *What is the Circulation of the blood?*

The Circulation of the blood is its movement through the body by which it is constantly passing to every organ and tissue.

317. *Describe the route ordinarily taken by the blood in circulation.*

Starting from the left ventricle of the heart, the blood passes out through the Aorta and its branches to the capillaries in the various tissues of the body; from the systemic capillaries it is collected into veins which return it through the Venæ Cavæ to the right auricle; from the right auricle it passes into the right ventricle; from the right ventricle it is forced out through the Pulmonary Artery and its branches to the capillaries in the lungs; from the pulmonary capillaries it is gathered into veins which return it through the Pulmonary Veins to the left auricle; and from the left auricle it passes to the left ventricle, the starting point.

318. *By what special route does some of the blood pass in making the circuit of the body?*

The blood which passes out from the left ventricle through the Aorta and its branches to the stomach, spleen, pancreas, and intestines, is collected from the capillaries of those organs through small veins into the Portal Vein, which carries it to the liver to be distributed in capillaries throughout the substance of that organ; collected again into veins, it passes through the Hepatic Veins and Inferior Vena Cava to the right auricle; from the right auricle it passes through the right ventricle, Pulmonary Artery, lung capillaries, Pulmonary Veins, left auricle, to the starting point in the left ventricle. This blood, in its round, passes through three sets of capillaries: in walls of intestines, in liver, and in lungs.

319. *What is the time required for the blood to complete the circuit of the body?*

From twenty to thirty pulsations, or, according to Flint, about twenty-three seconds.

520. *What are the "two circulations"?*

The passage of the blood through the body is called the "Systemic Circulation;" and that through the lungs, the "Pulmonic Circulation." The two together, however, form but one *circulation.*

521. *What changes take place in the blood in the systemic capillaries?*

Blood in passing through the capillaries gives up the oxygen carried by its cells and materials for the repair and renovation of its tissues, and gathers up a burden of ashes and waste products to be carried to some excretory organ. On account of the loss of oxygen and the acquisition of waste matter the color of the blood changes from bright red to dark purple.

522. *What changes take place in the blood in the lungs?*

In the lungs the blood throws off its load of carbonic acid gathered in the systemic capillaries, and takes a new supply of oxygen from the air cells. The color changes from dark purple to bright scarlet, principally on account of the action of the oxygen on the hæmoglobin in the red corpuscles.

523. *What are the propelling forces in the circulation of the blood?*

The principal propelling force in the circulation of the blood is that exerted by the contraction of the chambers of the heart. Auxiliary forces are gravity, mechanical compression of the veins, and the action of the respiratory apparatus.

524. *How does the heart act in propelling the blood?*

When the two auricles become charged with blood from the venæ cavæ and pulmonary veins respectively, they contract simultaneously, forcing the blood down into the expanding ventricles; immediately the ventricles contract and force the blood out to the lungs and body. After a brief pause these movements are repeated, producing the rhythmical *beat* of the heart.

525. *What are the movements of the heart called?*

The contraction of a chamber of the heart is called a *Systole;* and its expansion, a *Diastole.*

526. *What produces the "beating of the heart" felt upon the left side of the chest.*

It is produced by the systole of the ventricles, which throws the apex of the heart upward against the side of the chest.

527. *What is the Pulse?*

The Pulse is the throbbing of an artery, caused by the spurting of the blood as it is forced onward by successive cardiac impulses.

528. *What is "Heart Disease"?*

The term "heart disease" is applied to any organic or functional ailment of the heart, such as the thinning or thickening of its walls, the wasting away of its valves, irregularity in its action. etc. Most cases of so-called heart disease are brain affections.

529. *What are Aneurisms and Varicose Veins?*

Anuerisms are permanent dilations and enlargements of portions of arteries by a partial rupture of their coats: similar enlargements in veins are called Varicose Veins.

530. *How can the flow of blood from a wound be checked?*

If the blood be bright red and spurting in jets, it is from an artery, and can best be checked by a compress *between the wound and the heart;* but if it be dark purple and flowing in a steady stream, it is from a vein, and the compress must be on the vein *on the side of the wound away from the heart.* Lint, dry earth, or a simple bandage applied directly to a small wound will check the flow.

531. *What is Congestion?*

Congestion is an unusual accumulation of blood in some organ or part of the body, resulting from checking the natural flow of the blood through the tissues involved. *Blushing* is temporary congestion of the small blood vessels of the face and neck; and a "*Cold*" is congestion of the lungs and viscera.

532. *What is Inflammation?*

Inflammation is a congested state of the capillaries and smaller blood vessels in some part of the body, caused by some irritation which produces an unusual determination of blood to the part. It is characterized by redness, heat, swelling and pain.

533. *What causes Fainting?*

Fainting is caused by a lack of blood in the brain; therefore a person exhibiting a tendency to faint should be placed in a horizontal position, that the heart may more easily drive the blood to that organ.

HEART AND LUNGS

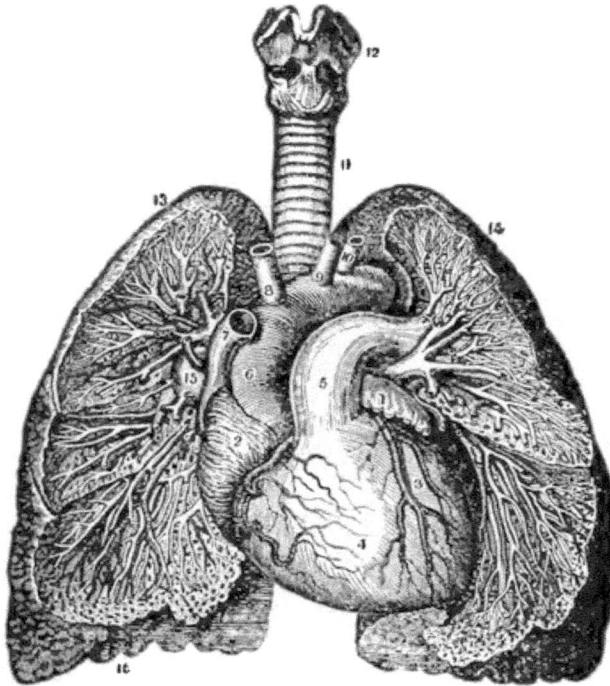

A View of the Bronchia and Blood-Vessels of the Lungs as shown by Dissection, as well as the relative Position of the Lungs to the Heart. 1, End of the Left Auricle of the Heart. 2, The Right Auricle. 3, The Left Ventricle with its Vessels. 4, The Right Vertricle with its Vessels. 5, Pulmonary Artery. 6, Arch of the Aorta. 7, Superior Vena Cava. 8, Arteria Innominata. 9, Left Primitive Carotid Artery. 10, Left Sub-Clavian Artery. 11, The Trachea. 12, The Larynx. 13, Upper Lobe of the Right Lung. 14, Upper Lobe of the Left Lung. 15, Trunk of the Right Pulmonary Artery. 16, Lower Lobes of the Lungs. The Distribution of the Bronchia and of the Arteries and Veins, as well as some of the Air-Cells of the Lungs, are also shown in this dissection.

RESPIRATORY SYSTEM.

Didactic Note—Special attention should be given to this System on account of its intimate relations to good health and the prevalence of disease resulting from the ignorant and heedless neglect of simple hygienic laws concerning it. Here, as elsewhere, the first requisite is a knowledge of the apparatus, both as seen with the unaided eye and as revealed by the microscope. The lungs, or "lights," of a hog, with the trachea attached, can be readily obtained for dissection in the presence of the class. The purification and oxygenation of the blood are attractive subjects for investigation, even for the youngest pupils. The instruction on the importance of pure air should be accompanied with warnings against "draughts." What may be called *fresh-air idiocy* has killed more people than all the wars in the world's history. Bad as are the effects of breathing air deficient in oxygen, the evils resulting from sitting in currents of air from open windows are ten times worse. Few people are injured permanently by rebreathing air in close rooms, while thousands die every year from "fresh air." The *prevention* of colds, nasal catarrh, consumption, etc., is undoubtedly the most important of all hygienic considerations; and teachers should give it the attention it demands. The instruction on voice and speech, which occurs incidentally in connection with this system, should be concreted in practical drills with the class.

SPECIAL OUTLINE.

1⁴ Respiratory System.   [See General Outline, page 8.]

  1⁵ Apparatus.

    1⁶ Air Passages.

      1⁷ Nasal Passages.

      2⁷ Mouth.

      3⁷ Pharynx.

      4⁷ Larynx.

      5⁷ Trachea.

      6⁷ Bronchi.

    2⁶ Lungs.

      1⁷ General description.

        1⁸ Location.

        2⁸ Form.

534. *What is the Respiratory System?*

The Respiratory System is a collection of organs whose function is the purification and oxygenation of the blood.

535. *Of what three parts does the respiratory apparatus consist?*

Air Passages, Lungs, and auxiliary apparatus.

536. *What are the Air Passages?*

The Air Passages are the various cavities and tubes through which the air passes to reach the lungs. They are the Nasal Passages, the Mouth, the Pharynx, the Larynx, the Trachea, and the Bronchi.

537. *Describe the Nasal Passages.*

The Nasal Passages are two large, irregular cavities in the nose, leading to the Pharynx. Their walls, formed of bone and cartilage, are covered by a delicate, moist membrane, supporting a growth of hairs which interlace across the passages.

538. *What are the Nostrils?*

The Nostrils, or *Nares,* are the anterior and posterior doorways of the nasal passages. Some anatomists apply the term to the passages themselves.

539. *Describe the mouth as an air passage.*

The mouth is a large, irregular passage, leading from the surface of the face at the lips to the Pharynx, into which it opens through the Fauces.

540. *Describe the Pharynx as an air passage.*

The Pharynx is a conical cavity, connecting both the nasal passages and the mouth with the Larynx. Food also passes through this cavity on its way from the mouth to the Œsophagus.

541. *Describe the Larynx.*

The Larynx is a conical box, composed of nine pieces of cartilage, situated beneath the base of the tongue, and in front of the upper portion of the œsophagus, and opening upward into the pharynx and downward into the trachea. It is lined by a very sensitive, delicate membrane.

542. *Name the cartilages of the Larynx.*

Thyroid, Cricoid, Epiglottis, two Arytenoid, two Cornicula Laryngis, and two Cuneiform.

543. *What is "Adam's Apple"?*

"Adam's Apple" is a prominence in front of the neck, formed by the Thyroid cartilage, the most important of the cartilages of the pharynx.

544. *Describe the Epiglottis.*

The Epiglottis is a thin, spoon-shaped plate of fibro-cartilage, which forms a sort of lid or trap-door for closing the entrance to the Larynx while food is passing through the pharynx.

545. *What are the Vocal Cords?*

The Vocal Cords are bands of yellow, elastic tissue, which stretch across the cavity of the larynx from the thyroid cartilage in front to the arytenoid cartilages behind. They are covered by a thin layer of closely-adherent mucous membrane.

546. *What is the Glottis?*

The Glottis is the slit-like opening between the vocal cords. It is v-shaped, with the angle in front, and varies in width with the tension of the cords.

547. *Describe the Trachea.*

The Trachea, or Windpipe, is a rigid, cylindrical tube, somewhat less than an inch in diameter and about four and a half inches in length, extending from the larynx down into the thorax, where it divides into the Bronchi. It is composed of yellow, elastic fibres, strengthened by from sixteen to twenty incomplete rings of cartilage, and lined by a mucous membrane.

548. *Describe the Bronchi.*

Opposite the third dorsal vertebra the trachea divides into two branches called the Bronchi, which extend diagonally outward to the lungs. The structure of these branches is the same as that of the trachea, except that the rings of cartilage are perfect.

549. *Describe the Lungs.*

The Lungs, the essential organs of respiration, are two large, pinkish-gray bodies, situated in the upper and lateral portions of the thorax, occupying about four-fifths of that cavity, and weighing together about forty-two ounces. The right lung is slightly larger than the left, owing to the inclination of the heart to the left side of the thorax.

550. *What is the Pleura?*

The Pleura is the investing membrane of the lungs. It is reflected upon the walls of the thorax so as to form about each lung a double sac, whose moist inner surfaces are brought into contact in a full inspiration.

551. *What are the different parts of the lungs called?*

The upper part of each lung is called the *Apex ;* the lower part, its *Base ;* and the part by which it is joined to the heart and trachea, its *Root.*

552. *Describe the structure of the Root of the lung.*

The Root of the lung is composed of the blood vessels leading from and to the heart, the bronchus, a plexus of nerves, lymphatics, and connective tissue—all wrapped in the reflection of the pleura.

553. *What are the Lobes of the lungs?*

Each lung is partially divided by deep fissures into *Lobes*, or smaller lung bodies, the right into three and the left into two. Each lobe is composed of still smaller masses of lung substances, called *Lobules.*

554. *Describe the structure of the lungs.*

The lungs are composed of a light, porous, spongy texture, called the Parenchyma, bound together by areolar tissue, and enclosed in a thin, transparent serous membrane.

555. *Describe the Parenchyma.*

The Parenchyma, or lung substance, is entirely made up of arteries, veins, lymphatics, bronchial tubes, and nerve filaments, which are arranged in minute, pear-shaped masses, called Lobules. These lobules are distinct from each other, but are closely bound together by connective tissue.

556. *Describe the air vessels of the lungs?*

A bronchus on entering the lung divides and subdivides, by twos, until the ultimate divisions are reached, which terminate in pear-shaped expansions, called *Infundibula.* On the walls of the Infundibula are minute pouches, called *Air Cells,* separate from each other, but all opening into the central cavity. It is estimated that the lungs contain over 600,000,000 of those cells.

557. *Describe the blood vessels of the lungs.*

The branches of the Pulmonary Artery upon entering the lungs divide again and again until they reach the capillaries, which form a dense net-work upon the walls of the infundibulæ and air cells ; from these capillaries arise minute veinlets, which unite to form the Pulmonary Veins, through which the blood returns to the heart.

558. *Of what does the auxiliary apparatus of the respiratory system consist?*

The " thoracic cage," the muscles of the thorax, and the diaphragm.

559. *Describe the Thoracic Cage as a part of the respiratory apparatus.*

The Thoracic Cage is a conical cavity, formed by the spinal column, the sternum, and the ribs. The ribs slope downward and forward in such a manner that raising their front extremities enlarges the cavity of the thorax and causes a current of air to flow inward; depressing them decreases the cavity and forces a current of air outward.

560. *What thoracic muscles aid in producing the respiratory movements?*

The *Levatores costarum* and *External intercostales* by their combined action raise the ribs and thus enlarge the thoracic cavity; the *Internal intercostales* and the *Triangularis sterni* draw the ribs downward and thus diminish the thoracic cavity.

561. *Describe the Diaphragm as a part of the respiratory apparatus.*

The Diaphragm is a muscular wall between the thorax and the abdomen. When at rest, it arches up into the thorax; when contracted, it is drawn down in the center and flattened, thus increasing the cavity of the thorax. It is the principal agent in producing inspiration.

562. *What are the physical characteristics of the Atmosphere?*

The Atmosphere, or air, is a colorless, elastic fluid surrounding the earth and flowing readily into all cavities under the pressure of gravitation. It expands under the action of heat, and holds vapors and gases in solution.

563. *What is the chemical composition of the atmosphere?*

The atmosphere is not a chemical compound, but a mere mechanical mixture of two gases, Oxygen and Nitrogen, in the ratio of 21 to 79, by volume. It also contains, at all times, small quantities of watery vapor and carbonic acid.

564. *Which of the two gases composing air is necessary to life?*

Oxygen; it is a necessary factor in the renovation of tissues and in the producing of heat. Nitrogen seems, so far as concerns animal life, to be valuable only as a means of diluting the otherwise too active oxygen.

565. *What are the two mechanical movements of respiration, or breathing?*

Inspiration and Expiration.

566. *Describe the process of inspiration.*

By depressing the diaphragm and elevating the sternum and anterior extremities of the ribs, the thoracic cavity is enlarged, the air contained in the lungs expands and becomes rarer, and external air, under the influence of gravity, rushes in to restore the equilibrium.

567. *Describe the process of Expiration.*

By the rising of the diaphragm into the thorax and the depressing of the ribs, the thoracic cavity is made smaller, and a portion of the air contained in the lungs is forced out.

568. *What effect has the elasticity of the lung structure upon respiration?*

The respiratory movements depend in part upon the elasticity of the lungs, which tend constantly to pull together, but are prevented by pressure of air within not balanced by pressure of air without in the unyielding, air-tight thorax.

569. *What do the terms Stationary Air and Tidal Air mean?*

The lungs always contain some air, which has been called by physiologists *Stationary Air*, while that which flows in to join it during an inspiration is called *Tidal Air*.

570. *What do the terms Residual Air, Supplemental Air, and Complemental Air mean?*

The air remaining in the lungs after the most violent expiration, is called *Residual Air ;* the difference between the Residual Air and the Stationary Air, during ordinary breathing, is called *Supplemental Air ;* the additional air which can be forced into the lungs and air passages by a violent effort after an ordinary inhalation, is called *Complemental Air.*

571. *What is the Vital Capacity?*

The *Vital Capacity* is the entire volume of air that can be taken into the lungs by the most violent inhalation after the most violent expiration. For the average healthy man it is about 225 cubic inches, while for the same person the amount taken in by an ordinary inhalation is only about 30 cubic inches.

572. *What is the average rate of breathing?*

Under ordinary circumstances a healthy person, sitting quietly and not conscious that his respirations are being counted, will breathe about *fifteen* times a minute.

573. *What amount of air is required for each person per day?*

Counting 30 cubic inches for each inspiration, and 15 inspirations per minute, the total amount required for each person in 24 hours is 374 cubic feet.

574. *What changes take place in the air in the lungs?*

The change in the air while in the lungs is fourfold : in temperature, in volume, in humidity, and in chemical composition. The first three are closely related : the air is warmed by the heat of the lungs, expanded by being warmed, and drinks in more moisture on account of its expansion.

575. *What are the chemical changes in the air in the lungs?*

Air taken into the lungs gives up oxygen to the blood and takes carbonic acid from the same source. Air, on entering the lungs, is, ignoring *aqueous vapor*, about 21 per cent. *Oxygen* and 79 per cent. *Nitrogen*, with a trace of *Carbonic Acid;* the same air on leaving the lungs is about 16 per cent. *Oxygen*, 80 per cent. *Nitrogen*, and 4 per cent. Carbonic Acid.

576. *Describe the process by which air reaches the ultimate divisions of the bronchial tubes.*

Since, after the most violent forced expiration, there still remains in the lungs about 100 cubic inches of air, it would be a natural inference that the pure air inhaled would drive this vitiated air forward into the smaller passages and cells, and that the pure air itself remaining in the bronchi and larger tubes would be forced out again by the following expiration, and no oxygen would reach the blood; but this is not the case; the *tidal air*, on coming into contact with the *stationary air*, rapidly mingles with it by the law of diffusion of gases, aided by a peculiar alternate expansion and contraction of the minute tubes in the lung substance.

577. *Describe the process by which air gives up oxygen and receives carbonic acid in the lungs.*

Air in the *infundibula* and *air cells* is separated from the blood in the surrounding plexus of capillaries by a very delicate membrane, through which oxygen passes to the blood and carbonic acid to the air, in accordance with physical and chemical laws. The membrane admits of a ready *transfusion* of gases, which are impelled to the passage principally by the law of *diffusion* already mentioned. *Chemical affinity* also has some influence in the action.

578. *What are the functions of the respiratory apparatus in expressing thought?*

The spoken language, as a means of communicating thought, is entirely the product of the respiratory apparatus. Air is expired through the respiratory passages in such a way as to produce sound, which, modified by change in the form of the passage, becomes *speech*.

579. *By what two processes are sounds produced by the respiratory apparatus?*

Vocalization and Whispering.

580. *What is Vocalization?*

Vocalization is producing sound by the vibrations of the *vocal cords*. The cords are stretched with more or less tension across

the opening into the larynx, and air in expiration is forced through
the slit between them and across their edges, causing them to
vibrate. The sound produced varies in *loudness*, dependent upon
the *force* of the expiration; it varies in *pitch*, dependent upon the
tension of the cords; and it also varies in *quality*, dependent upon
the *texture* of the cords.

581. *What is Whispering?*

Whispering is producing a rustling sound by driving the air
outward through the respiratory passages, without vocalization.
Whispering is puffing through the air tubes with the vocal cords
entirely relaxed.

582. *What is Articulation?*

Articulation is the process of modifying whispered and voiced
sounds by the action of the lips, teeth, tongue, palate, etc. The
modification is mainly effected by interrupting the current.

583. *What is Speech?*

Speech is articulated sound; *i. e.*, sound carved into words by
the lips, teeth, tongue, etc.

584. *What is the difference between Speech and Song?*

In Song the current of vibrating or rustling air is less broken
by the vocal organs of the mouth than in Speech. Song, like
Speech, may be either whispered or voiced.

585. *What is Whistling?*

Whistling is forcibly exhaling or inhaling through the contracted orifice made by puckering the lips. The sound produced
may be partially articulated by the lips, cheeks, tongue and teeth.

586. *What are Sighing and Yawning?*

These two processes are alike in that they produce unarticu-
lated, unvoiced sounds. In Sighing a prolonged, silent inspiration is followed by an audible expiration; in Yawning both inspiration and expiration are forced and audible.

587. *How are the sounds in Laughing and Crying produced?*

The sounds in Laughing and Crying are very similar, in many
instances identical; they are produced by short and rapid spas-
modic expirations, in which the sound is more or less vocalized.

588. *What are Coughing and Sneezing?*

In Coughing and Sneezing there is a full inspiration followed
by an explosive expiration, caused by first keeping the glottis
closed and then suddenly opening it to permit the exit of the
dammed up air. If the air passes out through the mouth it is a
Cough; if through the nose, a Sneeze.

589. *What is Stammering?*

Stammering is a nervous affection in which the proper action of the vocal organs is disturbed by twitching. Consciousness of the defect increases it. A full inspiration before speaking is the best preventive.

590. *What is Snoring?*

Snoring is the flapping to and fro of the soft palate in the divided air current flowing in or cut through both nose and mouth.

591. *What is Asphyxia?*

Asphyxia is suffocation, or "oxygen starvation." It results either from cutting off the air supply, or from breathing air deficient in oxygen.

592. *What are Pleurisy, Croup, Bronchitis and Pneumonia?*

These diseases are inflammations of different portions of the respiratory apparatus. *Pleurisy* is inflammation of the *pleura*, or investing membrane of the lungs; *Croup*, of the mucous membrane of the larynx and trachea; *Bronchitis*, of the mucous membrane of the bronchial tubes; and *Pneumonia*, of the lung substance.

593. *What is Consumption?*

Consumption is a disease of the lungs, in which tubercles form, break down and ulcerate. It is characterized by a gradual wasting away of the body under the distress of a continual cough.

594. *What is Nasal Catarrh?*

Nasal Catarrh is inflammation of the mucous membrane of the nasal passages. It is characterized by a chronic discharge of fluid—often offensive in odor.

EXCRETORY SYSTEM.

Didactic Note.—The extent and character of the instruction on the various organs and functions of the Excretory System must be determined by each teacher in view of his class and his own opinions of the propriety of such instruction. The kidneys of a sheep, obtained from the butcher's shop, may be used in teaching the location, form and structure of the human kidneys, to which they are similar; and the skull of some animal will show the manner in which the teeth are inserted in the alveolar process. A small microscope or magnifying glass will reveal new beauties in the structure of the skin. Give special attention to the care of the hair, nails and teeth; and make the instruction on bathing as simple and practical as possible.

SPECIAL OUTLINE.

5⁴ Excretory System. [See General Outline, page 8.]

  1⁵ Urinary Organs.

    1⁶ Apparatus.

      1⁷ Kidneys.

        1⁸ General description.

        2⁸ Structure.

          1⁹ Mechanical.

            1¹⁰ Capsule.

            2¹⁰ Kidney substance.

            3¹⁰ Central cavity.

          2⁹ Microscopic.

            1¹⁰ Of Cortical portion.

            2¹⁰ Of Medullary portion.

            3¹⁰ Urinary vessels.

            4¹⁰ Blood vessels.

    2⁷ Ureters.

    3⁷ Bladder.

    4⁷ Urethra.

    2⁶ Functions.

      1⁷ Matter excreted.

      2⁷ Effect on the blood.

  2⁵ Lungs [See Respiratory System].

  3⁵ Liver [See Digestive System]

4⁵ Skin.

  1⁶ Structure.

    1⁷ Cuticle: Epidermis.

    2⁷ Cutis: Dermis.

  2⁶ Appendages.

    1⁷ Hairs.

      1⁸ General description.

595. *What is the Excretory System?*

The Excretory System includes all those organs which have for their common function the removal of waste matters from the body, by separating them from the blood.

596. *What are the organs of the excretory system?*

Kidneys, Lungs, Liver, and Skin.

597. *What are the Urinary Organs?*

The essential urinary organs are the Kidneys; and the accessory organs are the Bladder, in which the secretion is collected, the Ureters, two tubes which convey the secretion from the kidneys to the bladder, and the Urethra, through which the contents of the bladder are discharged.

598. *Describe the Kidneys.*

The Kidneys are two dark red, bean-shaped bodies, about four inches long, two inches wide, and one inch thick, situated in the back part of the abdominal cavity, one on each side of the spinal column, opposite the last dorsal and first lumbar vertebræ.

599. *Describe the mechanical structure of the kidneys.*

A longitudinal section of a kidney, dividing it as a bean would split, exhibits, first, a thin, transparent covering, called the *Capsule;* second, the kidney substance, consisting of an outer *Cortical Portion* and an inner *Medullary Portion;* and third, a central cavity, called the *Sinus.*

600. *Describe the microscopic structure of the Cortical Portion of the kidney substance.*

The Cortical Portion is a reddish-brown, friable layer, about one-fifth of an inch in thickness, and constituting about three-fourths of the entire kidney substance. It is composed of little masses of tubules imbedded in connective tissue, and accompanied everywhere by blood vessels and lymphatics.

601. *Describe the microscropic structure of the Medullary Portion of the kidney substance.*

The Medullary Portion consists of straight tubules, surrounded by blood vessels, and arranged in conical masses, called the *Pyramids of Malpighi,* which project into the central cavity.

602. *Describe the urinary vessels of the kidneys.*

The *Uriniferous Tubules,* which constitute the greater part of the kidney substance, commence in small globular sacs in the cortical portion, and after a very tortuous course discharge their contents into the central cavity from the summits of the Malpighian Pyramids. About 1,000 of these minute tubes have their mouths on the apex of each of the eight to eighteen pyramids.

603. *Describe the blood vessels of the kidneys.*

The Renal Arteries carry blood into the kidneys through openings in their concave sides, called *Hiluses,* and divide and subdivide until the ultimate divisions are reached, which are coiled in little tufts, surrounded by the globular sacs of the urinary vessels; from these little tufts of capillaries the blood is collected into the *Renal Veins,* which carry it back into the circulation.

604. *Describe the Sinus of a kidney.*

The Sinus of a kidney consists of an oblong principal cavity, called the *Pelvis,* with two or three branching chambers, called *Infundibula.*

605. *What are the Ureters?*

The Ureters are two ducts about the size of a goose-quill and about 16 to 18 inches long, through which the urine passes from the kidneys to the bladder.

606. *What is the Bladder?*

The Bladder is the reservoir into which the continuously secreted urine is collected. It is an oval-shaped sac about five inches long and three inches wide, with a capacity of a pint or more.

607. *What is the Urethra?*

The Urethra is the tube through which the urine is discharged from the bladder.

608. *What is the composition of the renal secretion?*

The Urine, or renal secretion, consists of water holding in solution waste matters taken from the blood. The waste matters -

consist almost wholly of crystalline nitrogenous substances, called *Urea* and *Uric Acid*, and mineral salts.

609. *What effect has the action by the kidneys upon the blood?*

The chief work of the kidneys is the removal of nitrogenous waste from the blood; they also remove the small amount of carbonic acid that has accumulated since leaving the lungs, and the blood which leaves the kidneys is, therefore, the purest blood in the body.

610. *What are the functions of the Lungs as excretory organs?*

The chief function of the Lungs as excretory organs is the removal of carbonic acid from the blood; they also remove small quantities of other waste matters and a considerable amount of watery vapor.

611. *What are the excretory functions of the Liver?*

The nature of the secretion of the Liver, the Bile, is not well understood, but on account of the failure to connect it in any important way with the digestive process, it is thought to be largely an excrementious substance, containing *carbon* and *hydrogen.*

612. *What is the Spleen?*

The Spleen is a dark red, flattened, oval body, situated at the left end of the stomach, and usually classed as a secretory or excretory organ. Its use is not understood; but it is probably concerned in the production and renovation of blood corpuscles.

613. *What is the Skin?*

The Skin is the outer covering of the body, which closely invests all its parts externally, and entering the external orifices of its cavities becomes the delicate mucous membrane.

614. *Of what two layers is the skin composed?*

The skin is composed of two distinct layers: an outer, the *Cuticle*, or *Epidermis;* and an inner, the *Cutis vera*, or *Dermis.*

615. *Describe the structure of the Cuticle.*

The Cuticle, or "scarf skin," is composed of layers of small, flat cells, soft and moist within, and becoming dry, horny scales on the surface. It has no nerves or blood vessels, but its inner cells contain the coloring matter of the skin.

616. *Describe the structure of the Cutis vera, or true skin.*

The Cutis vera is a close net-work of fibrous tissue, in whose meshes nerves and blood vessels ramify. It also contains numerous small excretory glands.

617. *What are the appendages of the skin?*

Hairs, Nails, and Teeth.

618. *What are Hairs?*

Hairs are filaments, or thread-like bodies, growing upon the surface of the skin.

619. *Describe the mechanical structure of hairs.*

Hairs are mere modifications of the cuticle, or outer layer of the skin. Each hair consists of a bulbous *Root*, imbedded in a depression in the cuticle, called a *hair follicle;* a hollow, cylindrical *Shaft;* and a tapering *Point*.

620. *Describe the microscopic structure of the root of a hair.*

The root of a hair is a bulbous enlargement contained in a minute pit formed by the sinking of a portion of the cuticle down into the cutis. At the bottom of the little pit is a small projection over which a cup-like depression in the root of the hair fits. It is from this projection that the hair is developed, and when it is destroyed the hair no longer grows in that follicle.

621. *Describe the microscopic structure of the shaft of a hair.*

A transverse section of the shaft of a hair exhibits a thin, outer *cortical layer*, a thicker *fibrous layer*, and a *central cavity* containing a *medullary substance*.

622. *What gives hairs their color?*

The color of hairs is due to pigment granules contained in the cortical layer.

623. *How many hairs are there in the human head?*

A careful count of the hairs on several heads shows the average to be from 90,000 to 140,000, with from 1,000 to 1,200 to the square inch.

624. *How do hairs "stand on end"?*

In and around the hair follicles are little bundles of muscular fibres, called *Erector pili*, which acting involuntarily under fear cause the hair to stand up.

625. *What are the functions of hairs?*

Hairs are a protection against heat and cold. Those of the face guard the orifices of the sensory cavities against the entrance of dust and other foreign substances; the eyebrows shade the eyes from strong light and prevent perspiration from entering the orbit, and the eyelashes, hairs in the nostrils and ears and about the mouth have similar functions.

626. *What are Nails?*

Nails are plates of horny substance placed on the dorsal surfaces of the extremities of the fingers and toes. They are a modified form of the cuticle.

627. *Describe the mechanical structure of nails?*

Nails are convex on their outer surface and concave to the digits, and are imbedded in grooves made by folds of the skin, being attached along almost their entire length. The part projecting back into the fold of the skin is called the *Root*, and the exposed part is called the *Body;* the white, crescent-shaped part of the body, next to the root, is called the *Lunula*, and the part of the other extremity not attached is called the *Free edge*.

628. *How do nails grow?*

Nails grow from their roots, pushing the part already formed outward toward the free edge. It is estimated that it requires about five months for the thumb nail to grow its full length and about four times as long for the nail of the great toe to accomplish a similar growth.

629. *What are the functions of nails?*

Nails protect the ends of the fingers and toes, and add greatly to the mechanical perfection of the hand as an instrument in grasping firmly and in picking up small objects.

630. *What are the Teeth?*

The Teeth are hard, ivory-like appendages of the skin set in folds of the mucous membrane pushed deep into sockets in the alveolar processes of the maxillary bones. They are not bones, but true appendages of the skin, developed from the mucous membrane.

631. *How many teeth are there?*

There are thirty-two teeth in a "full set," arranged symmetrically in the upper and lower jaws.

632. *What are Temporary Teeth?*

The Temporary, or "Milk Teeth," cut through the gums in early childhood, at from three months to three years of age. They are but twenty in number, and are gradually replaced by the full set of *Permanent Teeth* between the ages of six and eight years.

633. *What are the parts of a tooth?*

Each tooth consists of three parts: the *Body*, projecting above the gum; the *Neck*, surrounded by the gum; and the *Root*, or *Fang*, imbedded in the alveolar process. The upper portion of the body is called the *Crown*.

634. *Describe the structure of a tooth ?*

A vertical, or longitudinal section of a tooth exhibits a solid portion enclosing a cavity. The solid portion is composed mainly of *Ivory* or *Dentine*, which is covered in the body of the tooth with a thin layer of *Enamel* and in the root by a layer of *Cortical Substance*, or cement; the cavity contains a soft, sensitive substance, called the *Dental Pulp.*

635. *What are the Gums?*

The Gums are dense, fibrous structures closely investing the alveolar processes and surrounding the necks of the teeth.

636. *What are the four classes of teeth with respect to form and use?*

Incisors, Canine, Bicuspids and Molars, which are arranged symmetrically in each half of each jaw, as follows, beginning at the middle in front: two Incisors, one Canine, two Bicuspids and three Molars.

637. *What are the functions of the teeth ?*

The teeth belong to the digestive system and are the chief organs of mastication: the incisors are used in nipping or biting the food; the Canine, in tearing it into pieces; and the Bicuspids and Molars, in crushing and pulverizing it.

638. *What two varieties of glands discharge their secretions upon the surface of the skin ?*

Sudoriferous and Sebaceous.

639. *What are Sudoriferous Glands ?*

Sudoriferous, or "Sweat Glands," are minute glands whose office is separating perspiration from the blood and discharging it upon the surface of the skin.

640. *Describe the structure of a sweat gland.*

A sweat gland is a minute tube coiled and knotted into a globular mass, and discharging through a spirally-coiled duct upon the surface of the skin.

641. *What are the " Pores " of the skin ?*

The "Pores" of the skin are the mouths, or outlets, of the sweat glands.

642. *What is Perspiration ?*

Perspiration, or "Sweat," the secretion of the sudoriferous glands, is a transparent, colorless liquid, salt to the taste and having a peculiar odor.

643. *What is Insensible Perspiration ?*

When sweat evaporates as fast as it is secreted, it is called

Insensible Perspiration, but when it dampens the skin, standing in drops on its surface, it is called Sensible Perspiration.

644. *What is the average daily amount of perspiration?*

The average amount of perspiration excreted in twenty-four hours has been determined by experiments to be about two pounds.

645. *Why does violent exercise increase the flow of perspiration?*

In violent exercise a greater quantity of blood is driven to the glands in a given time, and an increased secretion is the result.

646. *What are Sebaceous Glands?*

Sebaceous Glands are minute, oil-secreting glands which discharge their softening and protecting secretion upon the surface of the skin, usually in the hair follicles.

647. *What are the functions of the skin?*

The skin protects and supports the soft parts of the body, is the organ of touch, and is one of the chief excretory organs.

648. *What waste matters are excreted through the skin?*

The skin excretes in perspiration and exhalation watery vapor, carbonic acid, urea, common salt, etc.

649. *What effect has the excretory action of the skin upon the blood?*

The skin, like the kidneys and the lungs, tends in its proper action to purify the blood. If the perspiration is excessive, as it is in violent exercise, it lessens the normal amount of water in the blood and produces " systemic thirst."

650. *What is Dandruff?*

Dandruff is persistent dead cells of the cuticle, occurring in irregular patches on the head and easily brushed away.

651. *What are Freckles?*

Freckles are irregular deposits of coloring matter in the skin.

652. *What are Warts?*

Warts are tufted excrescences on the surface of the skin composed of aggregations of overgrown papillæ.

653. *What are corns?*

Corns are thickened portions of the cuticle similar in structure to the nails. They are produced by pressure and friction upon the feet and other parts of the body, and are cured only by removing the cause.

654. *How does hair "turn gray"?*

Hair turns gray or white by the disappearance of pigment granules from the medulla and the formation of air cells in the same substance. How this takes place is not known, but it does not mean that the hair is dead.

655. *Does hair ever "turn gray in a single night"?*

There are well authenticated cases of the turning gray in a few hours of the entire growth of hair upon a head; and the sudden appearance of single gray hairs and small patches of them is not infrequent.

656. *What are Blisters?*

Blisters are wounds caused by separating the cuticle from the cutis, by burning or otherwise; the cuticle may be entirely removed or simply pulled away so as to form a cavity in which a watery pus collects. Holding the blistered part in milk gives the quickest relief.

657. *What is Erysipelas?*

Erysipelas is an inflammation of the skin, characterized by redness, heat and swelling, and frequently by small eruptions on the surface, which dry up, producing bran-like scales.

658. *What is Toothache?*

Toothache is pain in the teeth caused by the decaying or breaking away of the solid portion so as to expose the dental pulp to the action of air and other foreign substances. Neuralgia of the face is often mistaken for toothache.

659. *What is the proper mode of caring for the teeth?*

The teeth should be carefully cleaned after each meal, using a stiff brush and water. Particles of food may be removed by soft wood toothpicks; quills or metal toothpicks should not be used.

660. *What are the Hygienic reasons for bathing the surface of the body?*

Bathing opens the pores of the skin and maintains its normal action in excretion; it also removes excreted solid matters and prevents their reabsorption in a putrid, disease-producing state.

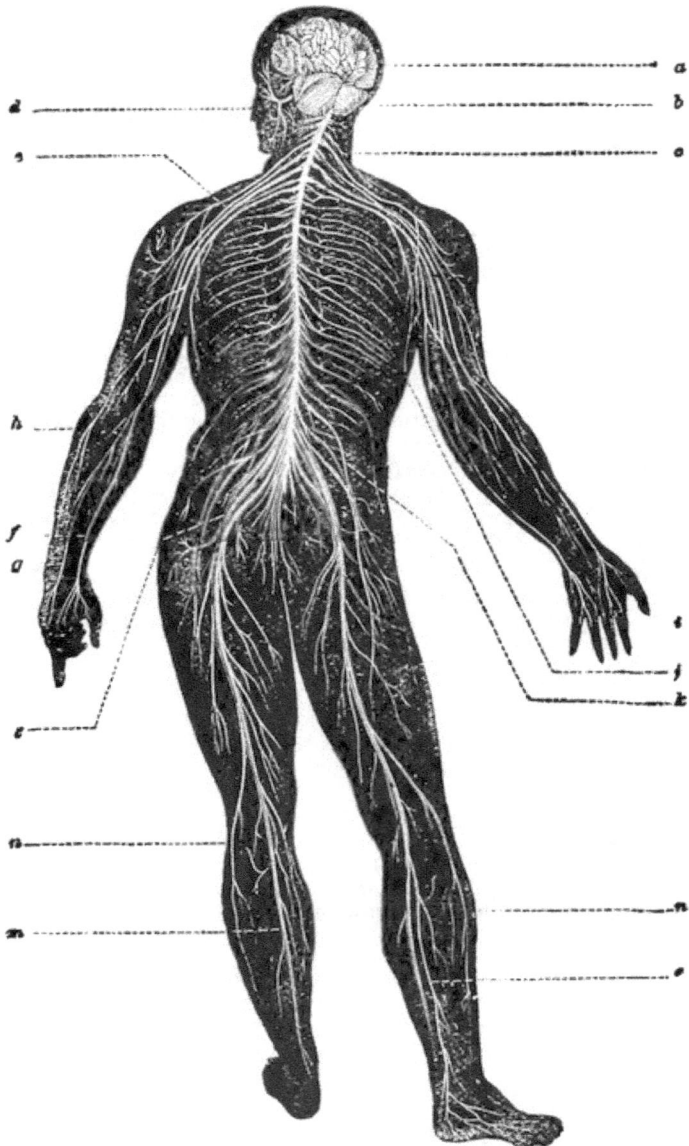

Nervous System.

a, Brain. *b*, Little Brain. *c*, Spinal Marrow. *d*, Facial Nerve. *e*, Brachial Plexus, caused by the union of several Nerves coming from the Spinal Marrow. *f*, Median Nerve. *g*, Cubital Nerve. *h*, Internal Cutaneous Nerve of the Arm. *i*, Radial and Musculo-Cutaneous Nerve of the Arm. *j*, Intercostal Nerves. *k*, Femoral Plexus. *l*, Sciatic Plexus. *m*, Tibial Nerve. *n*, External Peroneal Nerve ; *o*, External Saphenous Nerve.

CEREBRO-SPINAL NERVOUS SYSTEM.

Didactic Note.—The anatomy of this System can be readily shown by blackboard outlines and diagrams. Special attention should be given to the selection of the matter to be taught; the minute anatomy of the brain and the more abstruse theories of innervation should be excluded. Many questions will arise in the discussion, of which the teacher must be bold to say, "I don't know!" No egotistical, scholarly pride should cause him to assert as known that which is as yet a mere assumption of some over-confident, materialistic philosopher. The connection between the mind and the body is totally unknown, and all speculation upon that subject with children is improper. Impress the fact, however, that weakness and disease of the mind are usually associated with like states of the body, and urge a proper care of both.

SPECIAL OUTLINE.

1^4 Cerebro-Spinal System in General. [See General Outline, p. 8.]

  1^5 Apparatus.

    1^6 Cerebro-Spinal Axis.

      1^7 Brain: Encephalon.

        1^8 General description.

        2^8 Investing Membranes.

          1^9 Dura Mater.

          2^9 Arachnoid Membrane.

          3^9 Pia Mater.

        3^8 Divisions.

          1^9 Cerebrum.

            1^{10} General description.

            2^{10} Parts.

              1^{11} Hemispheres.

                1^{12} Right.

1^{13} Anterior Lobe.

2^{13} Middle Lobe.

3^{13} Posterior Lobe.

2^{12} Left.

  1^{13} Anterior Lobe.

  2^{13} Middle Lobe.

  3^{13} Posterior Lobe.

2^{11} Corpus Callosum.

3^{10} Convolutions of surface.

  1^{11} Eminences.

  2^{11} Depressions.

    1^{12} Fissures.

    2^{12} Sulci.

661. *What is the Nervous System?*

The Nervous System is a collection of organs which bring the various parts of the body under the control of the mind and into sympathetic relation with each other.

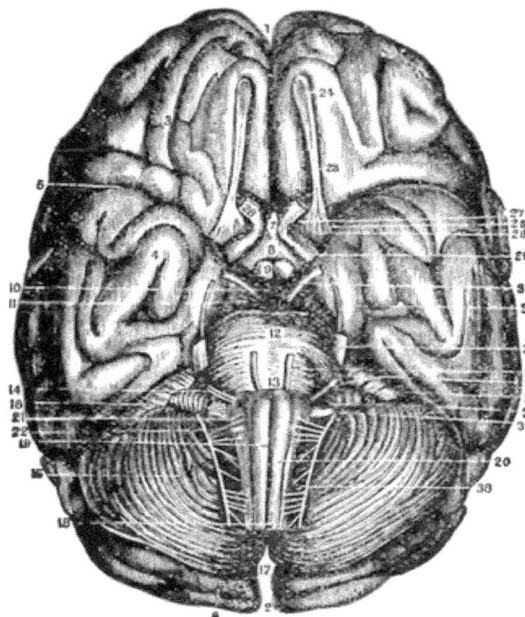

A View of the Base of the Cerebrum and Cerebellum, together with their Nerves. 1, Anterior Extremity of the Fissure of the Hemispheres of the Brain. 2, Posterior Extremity of the same Fissure. 3, The Anterior Lobes of the Cerebrum. 4, Its Middle Lobe. 5, The Fissure of Sylvius. 6, The Posterior Lobe of the Cerebrum. 7, The Point of the Infundibulum. 8, Its Body. 9, The Corpora Albicantia. 10, Cineritious Matter. 11, The Crura Cerebri. 12, The Pons Varolii. 13, The Top of the Medulla Oblongata. 14, Posterior Prolongation of the Pons Varolii. 15, Middle of the Cerebellum. 16, Anterior Part of the Cerebellum. 17, Its Posterior Part and the Fissure of its Hemispheres. 18, Superior Part of the Medulla Spinalis. 19, Middle Fissure of the Medulla Oblongata. 20, The Corpus Pyramidale. 21, The Corpus Restiforme. 22, The Corpus Olivare. 23, The Olfactory Nerve. 24, Its Bulb. 25, Its External Root. 26, Its Middle Root. 27, Its Internal Root. 28, The Optic Nerve beyond the Chiasm. 29, The Optic Nerve before the Chiasm. 30, The Motor Oculi, or Third Pair of Nerves. 31, The Fourth Pair, or Pathetic Nerves. 32, The Fifth Pair, or Trigeminus Nerves. 33, The Sixth Pair, or Motor Externus. 34, The Facial Nerve. 35, The Auditory—the two making the Seventh Pair. 36, 37, 38, The Eighth Pair of Nerves. (The Ninth Pair are not here seen.)

662. *What are the two great divisions of the nervous system?*

The Cerebro-Spinal and Sympathetic Systems.

663. *What are the divisions of the Cerebro-Spinal System?*

The Cerebro-Spinal System is usually considered in two divisions, the *System in General* and the *Organs of Special Sense.*

664. *What are the parts of the Cerebro-Spinal System in General?*

The Cerebro-Spinal Axis, Nerves, Ganglia, and Terminal Endorgans.

665. *What are the two parts of the Cerebro-Spinal Axis?*

The Brain and Spinal Cord.

666. *What is the Brain?*

The Brain, or Encephalon, is that part of the cerebro-spinal axis which is contained within the cranial cavity of the skull; it is a soft, egg-shaped body, weighing about fifty ounces, and constituting the chief ganglionic center of the nervous system.

667. *What three membranes invest the brain?*

The Dura Mater, Arachnoid Membrane, and Pia Mater.

668. *Describe the Dura Mater.*

The Dura Mater is a dense, inelastic membrane of white fibrous tissue, which lines the interior of the skull and sends down three strong processes for the support of the brain.

669. *Describe the Arachnoid Membrane.*

The Arachnoid Membrane is a delicate web of blended fibers of white and yellow connective tissue, which loosely invests the brain between the Dura Mater and the Pia Mater.

670. *Describe the Pia Mater.*

The Pia Mater is a vascular membrane which closely invests the surface of the brain, dipping down into all its fissures.

671. *What are the four parts of the brain?*

The Cerebrum, Cerebellum, Pons Varolii, and Medulla Oblongata.

672. *What is the Cerebrum?*

The Cerebrum is the principal part of the brain; it lies in the upper and front part of the cranial cavity, and is the seat of intelligence.

673. *What are the two great divisions of the Cerebrum?*

The Cerebrum is divided into the Right and Left Hemispheres by the Great Longitudinal Fissure extending along the middle line from the front to the back of the brain. The hemispheres are joined together at their base by a broad band of white nerve matter, called the Corpus Callosum.

674. *What are the Lobes of the brain?*

Each hemisphere of the brain is divided by deep fissures into three parts, called the Anterior, Middle and Posterior Lobes. The fissure which divides the Anterior from the Middle Lobe is called the *Fissure of Sylvius.*

675. *What are the Convolutions of the Cerebrum?*

The Convolutions of the Cerebrum are rounded folds on the surface of the brain, separated from each other by fissures about an inch in depth, called *Sulci.*

676. *Describe the structure of the Cerebrum.*

The Cerebrum is composed of two kinds of matter: a medullary, or white substance, constituting the principal part, and a cortical layer of gray matter, covering all its surface and dipping down into its fissures and sulci.

677. *What is the Cerebellum?*

The Cerebellum, or " Little Brain," is that portion of the brain which is situated in the lower and back part of the cranial cavity.

678. *What are the three parts of the Cerebellum?*

The Cerebellum is divided into two lateral *Hemispheres*, between which is a *Median Lobe.*

679. *Describe the structure of the Cerebellum.*

A vertical section through the inner third of either hemisphere of the Cerebellum exhibits a tree-like structure, called the Arbor Vitæ, the trunk being composed of white nerve substance and the branches of gray.

680. *What is the Corpus Dentatum?*

The Corpus Dentatum is a mass of gray matter imbedded in the central white matter of the Cerebellum.

681. *What is the Pons Varolii?*

The Pons Varolii is a central mass of gray and white fibres, constituting the bond of union of the different divisions of the brain; it joins the Cerebrum to the Medulla Oblongata below, and to the Cerebellum behind.

682. *Describe the structure of the Pons Varolii.*

The Pons Varolii is composed of alternate layers of transverse and longitudinal fibres intermixed with gray matter.

683. *What is the Medulla Oblongata?*

The Medulla Oblongata is the enlarged upper extremity of the spinal cord, situated within the cranial cavity.

684. *What are the parts of the Medulla Oblongata?*

The Medulla Oblongata is divided longitudinally by fissures into two symmetrical halves, each of which is subdivided in like manner into four columns, called, from before backwards, the Anterior Pyramid, Lateral Tract, Restiform Body, and Posterior Pyramid.

685. *What are the Olivary Bodies?*

The Olivary Bodies are two oval masses situated in the upper portions of the Lateral Tracts.

686. *Describe the structure of the Medulla Oblongata.*

The Medulla Oblongata, like other parts of the brain, consists of both gray and white matter, but in this division the gray substance is within and the white on the surface.

687. *What are the Ventricles of the brain?*

The Ventricles are chambers in substance of the brain formed by dilations of the minute central canal of the spinal cord on entering the skull. They are four in number, known as the First, Second, Third, and Fourth; and a Fifth is recognized by some anatomists.

688. *What is the Spinal Cord?*

The Spinal Cord is that part of the Cerebro-Spinal Axis which is contained within the vertebral canal of the spinal column.

689. *Describe the Spinal Cord.*

The Spinal Cord is a cylindrical mass of nerve matter extending downward from the brain to the lower border of the first lumbar vertebra, where it terminates in a slender process of gray substance, called *Filum Terminale.* It is about seventeen inches long and three-fourths of an inch in diameter.

690. *What are the two enlargements of the spinal cord?*

The spinal cord has two expansions; the Cervical Enlargement, extending from the third cervical to the first dorsal vertebra, and the Lumbar Enlargement, opposite the last dorsal vertebra.

691. *What are the fissures of the spinal cord?*

The spinal cord is creased longitudinally by the Anterior Median Fissure in front and the Posterior Median Fissure behind, partially dividing the cord into two symmetrical halves, which are joined in the middle line by a band called the *Commissure of the cord.* In each half, somewhat back of the middle, are grooves, called Lateral Fissures; and on each side of these are still smaller grooves, called Antero-lateral and Postero-lateral Sulci.

692. *What are the Columns of the spinal cord?*

The anterior and posterior median fissures and the lateral sulci divide each half of the spinal cord into three columns, called the Anterior, Lateral, and Posterior Columns.

693. *Describe the structure of the spinal cord.*

A transverse section of the spinal cord exhibits three membranes—the Dura Mater, Arachnoid Membrane, and Pia Mater—, enclosing a mass of white nerve substance, which contains a core of gray matter arranged in two crescent-shaped bodies.

694. *What are Nerves?*

Nerves are glistening white threads of nerve substance, extending from the cerebro-spinal axis and the various ganglia to all parts of the body.

695. *Describe the structure of a nerve.*

A transverse section of a nerve exhibits the *Neurilemma*, enclosing fatty material called the *White Substance of Schwann*, in which there is imbedded a transparent thread called the *Axis Cylinder*.

696. *What are the parts of a nerve?*

A nerve consists of a conductile thread and its modified extremities; the inner or central extremity is called its Origin, or Root, and the outer extremity, its Peripheral Termination.

697. *What are the two classes of nerves with respect to function?*

Nerve fibres which convey impressions from the various parts of the body to the nervous centers are called *Afferent* or *Sensory Nerves;* and those that convey impressions from the centers outward are called *Efferent* or *Motor Nerves.*

698. *What is the difference between afferent and efferent nerves?*

The principal difference between afferent and efferent nerves, so far as known at present, consists in difference in peripheral terminations.

699. *What are the two classes of nerves as to origin?*

Cranial, or those having their origins within the skull; and Spinal, or those having their origins along the spinal column.

700. *How are the Cranial Nerves classified?*

There are, according to the classification of Sömmerring, twelve pairs of cranial nerves: 1st, Olfactory; 2nd, Optic; 3rd, Motor oculi; 4th, Pathetic; 5th, Trifacial; 6th, Abducens; 7th,

Facial ; 8th, Auditory ; 9th, Glosso-pharyngeal ; 10th, Pneumo-gastric; 11th, Spinal Accessory ; and 12th, Hypoglossal. Of these the 1st, 2nd, 5th, 8th and 9th pairs are nerves of the organs of special sense, and should receive attention in connection with those organs.

701. *How are the Spinal Nerves classified?*

There are thirty-one pairs of spinal nerves leaving the spinal cord through the intervertebral foramina and grouped according to location into five classes, as follows: eight pairs *Cervical ;* twelve pairs *Dorsal ;* five pairs *Lumbar ;* five pairs *Sacral*, and one pair *Coccygeal.*

702. *Describe the origin or Root of a spinal nerve.*

A Spinal Nerve has two roots; an anterior or *motor* root arising from the anterior horn of the gray substance of the cord, and a posterior or *sensory* root arising from the posterior horn. The fibres of these roots intermingle and form a single nerve trunk which sends little branches throughout the tissues.

703. *How do nerves subdivide?*

Each nerve pursues its course separately from its origin to its terminus, but several nerves are bound together in a common sheath, called a *Funiculus*, which they branch out from.

704. *What is a Plexus of nerves?*

A Plexus of nerves is a net-work of nerves formed by the anastomosing of nerves from one funiculus to another.

705. *What are Ganglia?*

Ganglia are auxiliary nervous centers, similar in structure to the brain, though much less complex.

706. *Where are Ganglia situated?*

Ganglia are found at the posterior roots of the spinal nerves and at the junction of two or more nerves.

707. *How do nerves terminate?*

Nerves terminate in General and Special modes. The General mode is in delicate net-works of nerve filaments; motor nerves and some sensory nerves terminate in this way. The Special modes are called Peripheral End-organs.

708. *What are Peripheral End-organs?*

Peripheral End-organs are terminations of sensory nerves.

709. *What are the principal varieties of Peripheral End-organs?*

The End Bulbs of Krause, the Tactile Corpuscles of Wagner, and the Pacinian Corpuscles.

710. *What are the End Bulbs of Krause?*

The End Bulbs of Krause are minute nucleated capsules containing a soft homogeneous core into which the axis cylinder of the nerve fibre passes and terminates in a coiled mass.   They are found in the conjunctiva of the eye and the mucous membrane.

711. *What are the Tactile Corpuscles of Wagner?*

The Tactile Corpuscles of Wagner are minute corpuscles containing a soft structureless core in which the nerve fibres terminate by bulbous enlargements.   They are found in the papillæ of the skin of the fingers and toes.

712. *What are Pacinian Corpuscles?*

Pacinian Corpuscles are small masses composed of layers of cellular tissue arranged like the bulb of an onion and containing a minute watery core in which the nerve filament terminates in a bulbous knot.   They are found in the fingers and toes and beneath the skin generally.

713. *What are the functions of the nervous system?*

The two principal functions of the nervous system are the transmission of impressions from the various parts of the body to the brain for recognition and interpretation by the mind, and the conveying of messages from the brain and ganglia to the various organs of the body.   It controls the vital processes and brings the different parts of the body into sympathetic relation with each other.   It brings the mind into communication with the outer world through the Special Senses.

714. *What evidence is there that the Cerebrum is the seat of the mind?*

When the Cerebrum is small its possessor is an idiot; injury to the Cerebrum frequently produces idiocy or insanity; and a well developed cerebral structure is usually indicative of good mental development.   Intellectual activity is thought to deepen the sulci, increasing the cortical substance.

715. *What is the function of the Cerebellum?*

From experiments on living pigeons, it has been discovered that the principal function of the Cerebellum is the control of the voluntary muscles.

716. *What are the two classes of nerve centers?*

Nerve centers may be grouped into Reflex Centers and Centers of Consciousness.

717. *What are Reflex Nerve Centers?*

Reflex Nerve Centers are ganglia which receive impressions and control movements independent of consciousness, as when the burnt hand is snatched away before the consciousness of pain.

718. *What are Centers of Consciousness?*

Centers of Consciousness are centers of mind control in which impressions become sensations and emotions, and in which volitions originate.

719. *Where do we feel pain?*

Experiment has shown that the sensation of pain exists in the brain only.

720. *Why does striking the elbow cause the fingers to tingle?*

Compressing at the elbow the nerve leading to the hand produces a sensation of pain, which is interpreted as originating in the peripheral termination of the nerve in the third and fourth fingers.

721. *What is the rate of the transmission of a nervous impulse?*

A nervous impulse travels at the rate of somewhat more than 100 feet per second, which is very much slower than the transmission of sound.

722. *What is Sleep?*

Sleep is a cessation more or less complete of all forms of brain activity, during which the vital functions are performed in an imperfect and modified manner.

723. *What is Dreaming?*

Dreaming is activity of the mind in the semi-conscious state intervening between sound sleep and waking.

724. *What is Nightmare?*

Nightmare is a distressing dream occasioned by a disordered state of the digestive system.

725. *What is Neuralgia?*

Neuralgia is a diseased state of the nerves characterized by intermittent pains.

726. *What is Paralysis?*

Paralysis is the impairment or loss of sensation and the power of voluntary movement in a part of the body; it results usually from injury or disease of the spinal cord.

727. *What is the condition of a healthy mind?*

A healthy mind requires a healthy body, proper exercise, rest, cheerfulness, and control of the passions.

SPECIAL SENSES.

Didactic Note.—The structure of the eyeball should be taught
by blackboard drawings and the dissection of two or three beef's
or sheep's eyes in the presence of the class. Some knowledge of
optics is necessary to teach this subject well, and teachers who
have not studied physics should purchase an elementary text-
book and make such preparation as can be made by private study.
The anatomy of the ear is more difficult to teach than that of the
eye, but the general plan of the apparatus can be readily shown
on the blackboard. A few simple experiments, such as are given
in the common school text-books on Physiology, will add greatly
to the interest of the study.   A double convex lens may be used
to illustrate the action of the crystalline lens; simple tests for
color-blindness may be improvised; any simple musical instrument may be used in illustrating pitch, volume, and quality of
sound; try tasting an onion, holding the nose firmly and not inhaling; try salt, sweet and sour things on different parts of the
tongue; roll a marble between the crossed ends of the first and
second fingers, etc. Impress the importance of keeping these
door-ways of the senses, "the windows of the soul," open for the
reception of impressions of the external world, and show the
influence of bad habits upon the accuracy and reliability of sense
perceptions.

SPECIAL OUTLINE.

2^4 Organs of Special Sense.   [See General Outline, page 8.]

1^5 The Eye.
1^6 Apparatus.
1^7 Essential.
1^8 General description.
2^8 Structure.
1^9 Tunics.
1^{10} Sclerotic and Cornea.

2^{10} Choroid, Iris, Cil-
iary Processes.
3^{10} Retina.
2^9 Humors.
1^{10} Aqueous.
2^{10} Crystalline Lens.
3^{10} Vitreous.

728. *What are Sensations?*

Sensations are states of consciousness produced by impressions which originate in changes in the various parts of the body; these are conveyed to the brain along the sensory nerves and are interpreted by the mind.

729. *What are the two classes of sensations?*

Common and Special sensations.

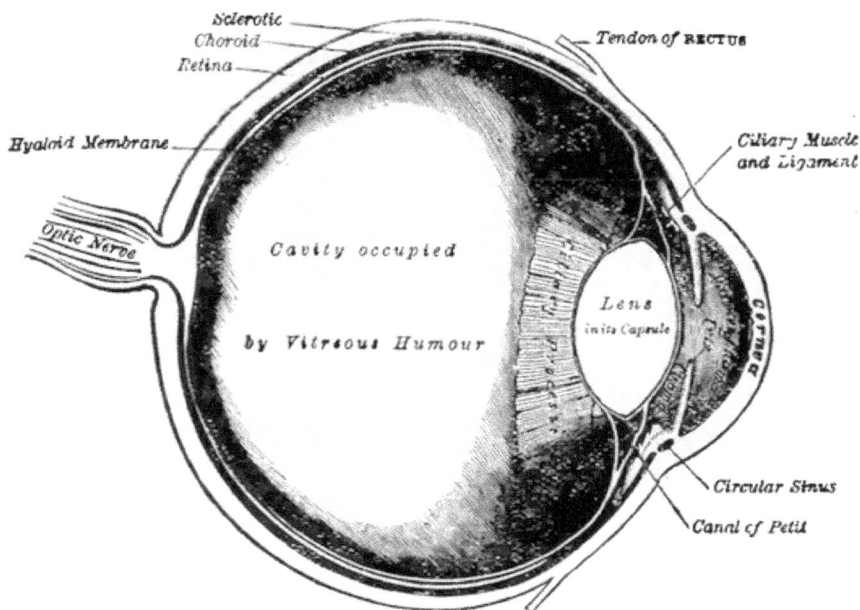

VERTICAL SECTION OF THE EYEBALL.

730. *What are Common sensations?*

Common sensations are perceptions of the simple physical properties and states; such as hardness, roughness, temperature, weight, hunger, thirst, weariness, etc.

731. *What are Special sensations?*

Special sensations are differentiated modifications of the common sensations resulting from impressions made on the organs of Special Sense.

732. *What are Organs of Special Sense?*

The Eye, Ear, Nose, Tongue, and Skin.

733. *What is the Eye?*

The Eye is the organ of sight.

734. *What are the essential parts of the Eye?*

The Eyeball and the Optic Nerve.

735. *Describe the Eyeball.*

The Eyeball is a spherical body about an inch in diameter, securely lodged in a bony cavity in the front part of the skull.

736. *What are the parts of the Eyeball?*

The Eyeball consists of three coats and three refracting media.

737. *What are the Coats or Tunics of the Eyeball?*

The coats of the eyeball named from without inward are, 1st, the Sclerotic and Cornea; 2nd, the Choroid, Ciliary Processes and Iris; 3rd, the Retina.

738. *Describe the Sclerotic Coat.*

The Sclerotic Coat is a tough, white, opaque, fibrous membrane covering the back and sides of the eyeball, and appearing between the eyelids as the *white* of the eye.

739. *Describe the Cornea.*

The Cornea is the transparent covering of the rounded eminence in front of the eyeball.

740. *Describe the Choroid Coat.*

The Choroid Coat is a vascular tunic consisting mainly of blood vessels and loose connective tissue in which are imbedded black pigment granules.

741. *What are the Ciliary Processes?*

The Ciliary Processes are plaits or folds, about sixty in number, in the choroid coat toward the front of the eyeball.

742. *What is the Iris?*

The Iris is a circular, perforated curtain composed of muscular tissue, nerves and connective tissue. It is of different colors in different persons.

743. *What is the Pupil?*

The Pupil is the circular apperature in the center of the iris, through which light enters the eye.

744. *What is the Retina?*

The Retina is a delicate membrane formed by the expansion of the optic nerve. It is a peripheral end-organ and is the essential part of the organ of sight.

745. *What are the Refracting Media or Humors of the Eye?*

The Refracting Media of the eye are the Aqueous humor, the Crystalline lens and the Vitreous Humor.

746. *Describe the Aqueous Humor.*

The Aqueous Humor is a watery fluid occupying the front part of the cavity of the eyeball between the cornea and crystalline lens.

747. *What is the Crystalline Lens?*

The Crystalline Lens is a transparent, colorless, double convex lens situated in front of the eyeball between the aqueous and vitreous humors.

748. *What is the Vitreous Humor?*

The Vitreous Humor is a transparent, jelly-like substance which fills the concavity of the retina, and constitutes about four-fifths of the entire eyeball.

749. *What are the accessory parts of the organ of sight?*

The Eyebrows, Eyelids, Conjunctiva and Lachrymal apparatus.

750. *What is the function of the eyebrows, eyelids and eyelashes?*

The function of the eyebrows, eyelids, and eyelashes is to protect the eye from too strong light and foreign substances.

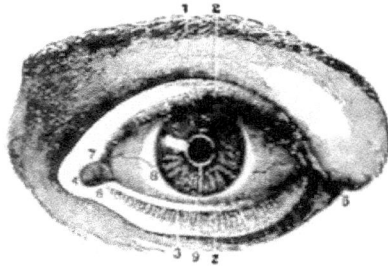

Front View of the Left Eye—moderately opened. 1, Supercilia. 2, Cilia of each Eyelid. 3, Inferior Palpebra. 4, Internal Conthus. 5, External Canthus. 6, Caruncula lacrymalis. 7, Plica Semilunaris. 8, Eyeball. 9, Pupil.

751. *What is the Conjunctiva?*

The Conjunctiva is a delicate, mucous membrane lining each eyelid and covering the front of the eyeball.

752. *What is the Lachrymal Apparatus?*

The Lachrymal apparatus consists of Tear Glands, Lachrymal Ducts and Nasal Ducts. Its function is the lubricating of the eyeball.

753. *How do we see?*

Rays of light from an object enter the eye through the pupil and are focused by the refracting media on the retina, producing nervous impressions, which, interpreted in the brain, give perceptions of the form, size, color, etc., of the object.

754. *What is "Color-blindness"?*

Color-blindness is inability to distinguish colors from each other. It is quite common, and should not cause distress, since similar imperfections are found in all the senses.

755. *What is "Nearsightedness"?*

Nearsightedness is inability to see distant objects distinctly. It results from too great convexity of the cornea and crystalline lens, and is corrected by the use of convex glasses.

756. *What are Spectacles?*

Spectacles are lenses set in frames and worn near the eyes for the purpose of strengthening them or correcting their defects.

757. *What is the "Blind spot"?*

The blind spot is the end of the optic nerves where it enters the back of the eyeball; rays of light falling on this spot produce no impression.

X   O

Hold the book directly in front of the face about four inches distant, and, closing the left eye, look steadily at the letter X with the right. Both letters are at first distinctly visible, but slowly move the book back from the face, keeping the right eye fixed on the letter X, and at the distance of about six inches the letter O will disappear as the rays of light reflected from it fall on the end of the optic nerve.

758. *Should young people wear glasses?*

Defects in the structure of the eyeball can frequently be cured or relieved by the use of proper lenses, and persons, whether young or old, should, on the discovery of such defects, be at once provided with spectacles selected by a competent oculist.

759. *What is the Ear?*

The Ear is the organ of hearing.

760. *What are the three divisions of the ear?*

The External, Middle, and Internal ear.

761. *What are the two parts of the External ear?*

The Pinna and Auditory Canal.

762. *What is the Pinna?*

The Pinna, or Auricle, is an irregularly folded sheet of cartilage attached to the side of the head for the purpose of collecting vibrations of sound into the funnel-shaped entrance of the auditory canal.

763. *Describe the structure of the Pinna.*

The Pinna is composed of a thin plate of yellow fibro-cartilage covered by the skin.

764. *What are the ridges and processes of the Pinna?*

The Helix, Antihelix ; Tragus, Antitragus, and Lobule.

765. *What are the depressions of the Pinna?*

The Fossa of the Helix, the Fossa of the Antihelix and the Concha.

766. *What is the Auditory Canal?*

The Auditory Canal, or *Meatus Auditorius Externus*, is

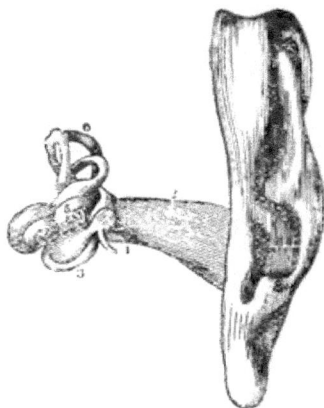

An Anterior View of the External Ear, as well as of the Meatus Auditorius, Labyrinth, etc. 1, The Opening into the Ear at the bottom of the Concha. 2, The Meatus Auditorius Externus or Cartilaginous Canal. 3, The Membrana Tympani stretched upon its Ring. 4, The Malleus. 5, The Stapes. 6, The Labyrinth.

an irregular, cylindrical tube, about an inch and a quarter long, which extends obliquely inward and forward into the *petrous portion* of the temporal bone.

767. *Describe the Tympanum or Middle Ear?*

The Tympanum is an irregular cavity, about the size of a small cherry, situated in the petrous portion of the temporal bone. It is filled with air, and traversed by a chain of little bones.

768. *What is the Membrana Tympani?*

The Membrana Tympani, or "Drum of the Ear," is a thin, semi-transparent membrane, situated at the inner extremity of the auditory canal, and separating it from the middle ear.

769. *What are the Auditory Ossicles?*

The Auditory Ossicles, or "Little Bones of the Ear," are a chain of small bones, the Malleus, Incus, and Stapes, which traverse the middle ear from the membrana tympani to the inner ear. They receive their names from a slight resemblance to the hammer, anvil and stirrup, respectively.

770. *What are the openings in the walls of the Ear ?*

There are three principal openings in the walls of the middle ear ; one into the Eustachian tube leading to the Pharynx ; the Fenestra Ovalis, a kidney-shaped opening, leading to the vestibule of the inner ear ; and the Fenestra Rotunda, a circular opening leading to the cochlea.

771. *Describe the Labyrinth, or Inner Ear ?*

The Labyrinth is a complex apparatus consisting of three parts, the *Vestibule, Semicircular Canals* and *Cochlea,* formed by a series of cavities in the petrous portion of the temporal bone.

772. *What is the Vestibule of the Inner Ear ?*

The Vestibule is the lobby or ante-chamber of the inner ear, communicating between its parts, and with the middle ear through the fenestra ovalis.

773. *What are the Semicircular Canals ?*

The Semicircular Canals are three channels, about one-twentieth of an inch in diameter, arching out into the bone separately from the vestibule, and returning to it after completing the greater parts of circles.

774. *What is the Cochlea ?*

The Cochlea is a spiral canal, having some resemblance to the common snail shell, extending from the vestibule into the substance of the temporal bone.

775. *How does the Auditory nerve reach the inner ear ?*

The Auditory nerve reaches the inner ear through the *Internal Auditory Meatus,* dividing at the bottom of that cavity into two branches, one leading to the vestibule and semicircular canals, and the other to the cochlea. The ultimate divisions and terminations of these branches are uncertain.

776. *How do we hear ?*

Sound waves collected in the external ear strike against the membrana tympani, which vibrates like the head of a drum, and sends the waves across the middle ear along the chains of little bones to the labyrinth, where they produce impressions upon the end-organs of the auditory nerve, which in the brain are interpreted as volume, pitch, quality, etc.

777. *What is the function of the External Ear?*

The external auditory canal, with its flared expansion, the pinna, is simply designed to collect and focus sound waves upon the membrana tympani.

778. *What are the functions of the Auditory Ossicles?*

The principal function of the little bones of the ear is the transmission of sound across the cavity of the middle ear, but they have also a curious secondary function: the handle of the malleus, in contact with the membrana tympani, acts in the same manner as the "dampers" on the strings of a piano, to check the vibrations of the membrane when the aerial vibrations cease.

779. *What parts of the ear are thought to determine the different properties of sound?*

Experiments and the study of the structure of the ear have led to the belief that the *volume* of sound is determined by the *tympanum*, the *pitch* by the *cochlea*, and the *quality* by the *semicircular canals*.

780. *What is the function of the Eustachian Tube?*

The Eustachian Tube connects the middle ear through the pharynx with the external ear, and thus equalizes the pressure on the inner and outer sides of the membrana tympani, preventing injury to the membrane and securing accuracy in its vibrations.

781. *What is Deafness?*

Deafness is inability to hear or distinguish sounds, resulting from defects in the parts of the ear or from paralysis of the auditory nerve.

782. *What care does the ear require?*

The ear is the most delicate organ of the body, and demands the greatest care. Hard substances or cold water should not be allowed to enter the ear. Hardened accumulations of wax can usually be readily removed by a little tepid water and a soft cloth.

783. *Describe the structure of the Nose.*

The Nose is a triangular body projecting from the middle of the face. Its walls are formed of plates of bone and cartilage, and its cavity is divided into large, irregular chambers by a vertical wall or septum.

784. *What is the Schneiderian Membrane?*

The Schneiderian Membrane is the mucous membrane lining the nasal fossæ; in its delicate structure are the terminal end-organs of the Olfactory nerve.

785. *How do we smell?*

Minute particles of the object smelled, floating in the air, are drawn into the nasal cavities by inhalation, and striking against the terminations of the nerve filaments produce impressions which are interpreted in the brain as odors.

786. *Why does "a cold in the head" temporarily, wholly or partially, destroy the sense of smell?*

The inflamed mucous membrane is less sensitive, and the passage of the current of air is impeded by collections of nasal excretion.

787. *What care does the nose require?*

Acuteness of smell, as well as purity of voice and common decency, requires that at least the entrance of the nasal cavities should be kept free from accumulations of nasal excretious.

788. *What is the Tongue?*

The Tongue is the organ of taste.

789. *What are the parts of the tongue?*

The Root, Lip, Dorsum, and Edges.

790. *Describe the structure of the tongue.*

The tongue is a mass of voluntary muscular fibres covered by a mucous membrane in which are the end-organs of the nerves of taste.

791. *What are the three classes of the papillæ of taste?*

Circumvallate, Fungiform, and Filiform Papillæ.

792. *Describe the Circumvallate Papillæ.*

The Circumvallate Papillæ, the largest of the three varieties, are flattened projections in little cup-like depressions, arranged in a v-shaped line on the back part of the dorsum of the tongue.

793. *Describe the Fungiform Papillæ.*

The Fungiform Papillæ are little deep-red, club-shaped eminences, situated on the tip and edges of the tongue and scattered sparsely over its dorsum.

794. *Describe the Filiform Papillæ.*

The Filiform Papillæ are very minute, brush-like processes which cover thickly the anterior two-thirds of the dorsum of the tongue.

795. *How do we taste?*

Minute particles of substances are brought into contact with the papillæ of taste, producing impressions which are interpreted in the brain as savors.

796. *What are the specific functions of the different varieties of papillæ?*

Experiments have shown that bitter and salt substances produce more impression on the back part of the tongue in the region of the circumvallate papillæ; sweet and sour substances, on the tip and edges; and pungent substances on the dorsum.

797. *What care does the organ of taste require?*

Acuteness in taste depends upon maintaining a healthy condition of the mucous membrane of the tongue, by avoiding substances which tend to destroy the papillæ and deaden the nerve filaments, as tobacco, strong condiments, etc.

798. *What is a "furred tongue"?*

A "furred tongue" is a coated condition of the dorsum of that organ, resulting usually from a disordered condition of the stomach. The "coating" consists of mucous epithelium cells, and a growth of microscopic organisms called *bacteria.*

799. *What is the organ of touch?*

The Skin, which is also a protective, supporting and excretory organ.

800. *Describe the papillæ of touch.*

The papillæ of touch are minute, conical eminences on the surface of the cuticle, about $\frac{1}{100}$ of an inch in height and $\frac{1}{250}$ of an inch in diameter, in which are found Pacinian Corpuscles and other varieties of terminal end-organs. They are most numerous on the tips of the fingers, which are specially the instruments of touch.

801. *What properties are discovered by touch?*

Through the sense of touch we determine the properties of hardness, roughness, etc., and their opposites.

802. *What is the distinction between feeling and touching?*

Feeling is general and *passive*, while touch is special and *active;* feeling is the general consciousness of physical states, while touching is a definite activity discovering specific properties in external bodies.

803. *What care does the organ of touch require?*

Acuteness of touch depends upon cleanliness of the skin and freedom from calluses. This sense is also rapidly deadened by the use of alcoholic stimulants and tobacco, and by various forms of dissipation and self-abuse.

SYMPATHETIC NERVOUS SYSTEM.

Didactic Note.—This division of the Nervous System is usually too much neglected by teachers. The general study of its apparatus is not difficult to understand; and its importance demands that it receive a fair share of attention. Show how in the reflex action which controls the vital processes the ganglia act as little brains; and explain the wisdom of the provision which places these life functions beyond the control of the will, thus freeing the mind from their constant care, and securing them against capricious action. The sympathetic relations of the vital organs may be made the subject of beautiful lessons, the materials for which may be drawn from the experiences and observations of the pupils themselves.

SPECIAL OUTLINE

2³ Sympathetic System.   [See General Outline, page 8.]
  1⁴ Structure.
    1⁵ Double Gangliated Cord.
      1⁶ Structure.
        1⁷ Cords.
        2⁷ Ganglia.
      2⁶ Divisions.
        1⁷ Cervical.
        2⁷ Dorsal.
        3⁷ Lumbar.
        4⁷ Sacral.
        5⁷ Coccygeal.
    2⁵ Plexuses.
      1⁶ Cardaic.

2⁶ Epigastric.
3⁶ Hypogastric.
3⁵ Subordinate Ganglionic
    Centers.
4⁵ Nerves.
  1⁶ Communicating.
  2⁶ Distributory.
2⁴ Functions.
  1⁵ Reflex control of visceral
    organs.
  2⁵ Co-ordination of vital ac-
    tivities.

804. *What is the Sympathetic Nervous System ?*
The Sympathetic Nervous System is a somewhat independent auxiliary division of the Sensory apparatus, situated in the

thoracic, abdominal, and pelvic cavities, and consisting of a number of ganglionic centers with connecting and radiating nerves.

805. *By what other names is the Sympathetic System known?*

Ganglionic System and System of Organic Life.

806. *Describe the apparatus of the Sympathetic System.*

The Sympathetic System consists of a central double chain of ganglia from which branch numerous nerves, forming three great plexuses, and distributing fibres to all the thoracic, abdominal, and pelvic viscera.

807. *Describe the ganglionic center of the sympathetic system.*

The center of the sympathetic system is two rows of ganglia extending through the thoracic, abdominal, and pelvic cavities, and situated in front and on either side of the spinal column. There are forty-nine of these ganglia, viz.: twenty-four pairs, and an odd one in front of the coccyx common to both chains.

808. *What are the five divisions of the double gangliated cord?*

The different portions of the gangliated cord, named from the regions in which they are found, are, beginning above, the Cervical, Dorsal, Lumbar, Sacral, and Coccygeal.

809. *What are the three great gangliated plexuses of the sympathetic system?*

Cardiac, Epigastric, and Hypogastric Plexuses.

810. *Describe the Cardiac Plexus.*

The Cardiac Plexus is a net-work of nerves beaded with ganglia, situated on the dorsal side of the heart.

811. *What are the two divisions of the Cardiac Plexus?*

The Superficial Cardiac Plexus lying beneath the arch of the aorta, and the Deep Cardiac Plexus lying behind the arch of the aorta and in front of the bifurcation of the trachea.

812. *Describe the Epigastric Plexus.*

The Epigastric or Solar Plexus is a dense net-work of nerves and ganglia situated behind the stomach and in front of the aorta.

813. *What are the Semilunar Ganglia?*

The Semilunar Ganglia are two irregular ganglionic masses situated on each side of the solar plexus, formed by the aggregation of smaller ganglia. They are the largest ganglia of the body.

814. *Describe the Hypogastric Plexus.*

The Hypogastric Plexus is a net-work of nerve filaments, without ganglia, situated in front of the promontory of the sacrum, between the two common iliac arteries, and sending nerves to all the pelvic viscera.

815. *What are the subordinate centers of the sympathetic system?*

Besides the central ganglionic chain and the three great plexuses, there are numerous minute ganglia which form additional centers for the origin of nerve fibres for supplying the separate viscera.

816. *What are the two classes of nerves of the sympathetic system?*

The Communicating, joining the various ganglia to each other and to the spinal nerves; and the Distributory, extending from the centers to the viscera.

817. *What are the functions of the sympathetic system?*

The sympathetic system controls by reflex action the vital functions of the thoracic and abdominal viscera, and brings into sympathetic relation the vital organs in such a manner that when one part suffers injury or distress all the others are more or less affected.

818. *Why is the sympathetic system called the System of Organic Life?*

On account of its intimate relations to the nutritive process, and because it is the sole nervous structure in the lowest forms of animal life, this system has been called the "System of Organic Life."

ALCOHOL AND NARCOTICS.

Didactic Note.—The importance of this subject and the prominence given it by legislative enactment demand that teachers make special preparation for teaching it. The instruction to be given is practically three-fold: 1st, concerning the nature and properties of the drugs themselves; 2nd, concerning their effects on the material organism, on its functional activities, and on the intellectual and moral tone of the individual; and 3rd, concerning the treatment, **both** curative and **preventive**. **In** teaching the properties of the drugs, the point to **be insisted** on is that they are *poisonous*, no matter in what **forms or in what** quantities they are taken. Concerning their effects, the **testimony of** medical experts **and the** observations of the pupils **themselves can be** confidently appealed to; **no one** can use any of these poisons in **any form without losing in** physical health, functional **accuracy and** vigor, and intellectual and moral power. The instruction on **the** treatment of the diseased **states must** manifestly be largely **on the preventive** phase of it; **but here and there a boy will** be found who, under the influence of home **surroundings and example,** has acquired the **tobacco** habit, with such the **conscientious** teacher will **labor earnestly for the** cure. **In the treatment of** tobacco, **which, on account of its more general and seemingly** respectable **use, many deny a place in the list of poisons, care** must be **used to prevent arousing by extreme utterances, even of** the truth, **a home antagonism which will largely undo all the** work of the school-room. **Great tact and patience with the** opinions of others are **required in such instruction. "Let not your good be evil spoken of;"** *but make no compromise with evil.* Numerous simple experiments are **given in the later text-books,** and others will **suggest themselves to the wide-awake teacher. Make the work** simple and practical, and **fix each lesson in the** memory by illustrations and examples from nature and life. No greater responsibility rests **upon teachers than that in connection** with this subject.

ANALYTIC OUTLINE.

1¹ Alcohol and Alcoholism.
  1² Alcohol.
    1³ Chemical composition.
    2⁴ Formation.
     1⁴ Sources.
      1⁵ Fruits.
       1⁶ Grapes, etc.

2⁶ Apples, etc.
2⁵ Grains.
  1⁶ Corn, etc.
  2⁶ Barley, etc.
3⁵ Potatoes, etc.
4⁵ Molasses, etc.
2¹ Processes.

1^5 Fermentation.
2^5 Distillation.
3^5 Rectification.
4^5 Brewing.
3^3 Properties.
  1^4 Solvent.
  2^4 Volatile.
  3^4 Antiseptic.
  4^4 Inflammable.
  5^4 Poisonous.
  6^4 Absorbent.
4^3 Forms.
  1^4 Fermented.
    1^5 From grape sugar.
      1^6 Ciders.
      2^6 Wines.
    2^5 From starch.
      1^6 Beer.
      2^6 Ale.
      3^6 Porter.
      4^6 Stout.
  2^4 Distilled.
    1^5 Brandy.
    2^5 Gin.
    3^5 Whisky.
    4^5 Rum.
5^3 Adulterations.
  1^4 Water, for profit.
  2^4 Drugs, for deception.
    1^5 Coculus Indicus.
    2^5 Etc.
6^3 Uses.
  1^4 As a food ?
  2^4 As a medicine ?
  3^4 As a luxury ?
  4^4 As a poison !
2^2 Alcoholism.

1^3 Forms.
  1^4 Chronic.
  2^4 Acute.
2^3 Causes.
  1^4 Predisposing influences.
  2^4 Exciting causes.
3^3 Pathology.
  1^4 Morbid anatomy.
  2^4 Functional derange-
    ment.
  3^4 Psychical derangement.
4^3 Diagnosis.
5^3 Prognosis.
6^3 Treatment.
  1^4 Curative.
    1^5 In acute stages.
    2^5 In chronic cases.
  2^4 Preventive.
    1^5 By personal resolution
    2^5 By state control.
2^1 Tobacco and Nicotinism.
1^2 Tobacco.
  1^3 Cultivation.
  2^3 Preparation.
  3^3 Adulteration.
2^2 Tobacco poisoning.
  1^3 Causes.
    1^4 Chewing.
    2^4 Smoking.
    3^4 Snuffing.
  2^3 Pathology.
    1^4 Morbid anatomy.
    2^4 Functional derange-
      ment.
    3^4 Psychical derangement.
  3^3 Treatment.
    1^4 Curative.

2^4 Preventive.
3^1 Opium and the Opium Habit.
  1^2 Opium.
    1^3 Source.
    2^3 Preparation.
    3^3 Forms.
  2^2 Opium poisoning.
    1^3 Causes.
      1^4 Eating opium.
      2^4 Injecting morphine hypodermically.
    2^3 Pathology.
      1^4 Morbid anatomy.
      2^4 Functional derangement.
      3^4 Psychical derangement.
    3^3 Treatment.
      1^4 Curative.
      2^4 Preventive.
4^1 Chloral and the Chloral Habit.
  1^2 Chloral.
    1^3 Source.
2^3 Preparation.
2^2 Chloral poisoning.
  1^3 Causes.
  2^3 Pathology.
    1^4 Morbid anatomy.
    2^4 Functional derangement.
    3^4 Psychical derangement.
  3^3 Treatment.
5^1 Cocaine and the Cocaine Habit.
  1^2 Cocaine.
    1^3 Source.
    2^3 Preparation.
  2^2 Cocaine poisoning.
    1^3 Causes.
    2^3 Pathology.
      1^4 Morbid anatomy.
      2^4 Functional derangement.
      3^4 Pschyical derangement.
    3^3 Treatment.

---

819. *What is Alcohol?*

Alcohol is the intoxicating principle of fermented liquors.

820. *What is the chemical composition of alcohol?*

In 100 parts of alcohol, 52.67 are carbon, 12.90 are hydrogen, 34.43 are oxygen.

821. *From what sources is alcohol obtained?*

Alcohol is obtained from the sweet juices of fruits and succulent vegetables, and from the starch of grain and root plants.

822. *How is alcohol formed?*

Alcohol is formed by the process of fermentation induced by yeast upon grape sugar.

823. *What is Fermentation?*

Fermentation is the process of decomposition or decay of sugar and starch.

824. *What is Yeast?*

Yeast is a microscopic fungus by whose growth the process of fermentation is accomplished.

825. *What is Distillation?*

Distillation is the process of separating the alcohol from water and other substances formed by the process of fermentation. Alcohol is *made* by fermentation, *separated* by distillation.

826. *Describe the process of distillation.*

Distillation is the process of purifying a liquid by heating it until it passes off in steam, and then condensing the steam into a liquid again in cold vessels.

827. *Describe the alcohol still.*

An alcohol still consists of a copper retort from which leads a coiled pipe passing through a large cask of cold water.

828. *How is alcohol distilled.*

Alcohol vaporizes at 172°, while water requires 212° of heat; therefore if alcohol mixed with water is heated in the still, the alcohol passes off first in steam, leaving the unvaporized water behind.

829. *What is Rectification?*

Rectification is the re-distillation of distilled liquors for the purpose of eliminating water and other impurities.

830. *What is Brewing?*

Brewing is the process of fermentation by which beer and similar alcoholic drinks are formed from barley and other grains.

831. *Describe the process of brewing.*

The barley is first dampened and sprouted, to turn the starch into sugar; then dried, to stop the growth of the sprouts so as to save the sugar; then ground and steeped in warm water, to dissolve the sugar; then boiled with hops and cooled in large vats; and then yeast is added to induce fermentation, which makes the alcoholic element of the beer.

832. *What is Malt?*

Malt is barley in the second stage of the beer formation, viz.: when it has been sprouted and dried.

833. *What is Sweet Wort?*

Sweet Wort is barley in the third stage of beer formation, viz.: after it has been sprouted, dried and "mashed" by grinding and steeping.

834. *What are the most important properties of alcohol?*

Alcohol is solvent, volatile, antiseptic, inflammable, poisonous, and has a strong affinity for water.

835. *What are the forms of the alcohol of commerce?*

Alcohol is sold in various degrees of purity, from the "Spirits of Wine," 90 per cent. strong, to the weakest grades of beer.

836. *What are the two classes of alcoholic drinks.*

Fermented and distilled.

837. *What are the two classes of fermented drinks?*

Those from sugar and those from starch.

838. *What are the two classes of alcoholic drinks obtained from sugar?*

Ciders and Wines.

839. *What are Ciders?*

Ciders are drinks made from apples and kindred fruits, by crushing the fruit and pressing out the juice.

840. *What are Wines?*

Wines are drinks made from grapes by crushing them and pressing out the juice.

841. *How is the alcohol of ciders and wines formed?*

The alcohol of ciders is the result of fermentation or decay of the grape sugar in the juices of the fruits.

842. *What are fermented drinks made from starch called?*

Malted or Brewed Drinks.

843. *What are the principal malted drinks?*

Beer, Ale, Porter, and Stout.

844. *What are the principal distilled drinks?*

Brandy, Gin, Whisky, and Rum.

845. *What are Brandies?*

Brandies are distilled wines and ciders; from grape wine is made grape brandy; from apple cider, apple brandy; from peach cider, peach brandy; etc.

846. *What is Gin?*

Gin is distilled beer.

847. *What is Whisky?*

Whisky is a distilled liquor obtained from the fermentation of corn, rye, potatoes, etc.

848. *What is Rum?*

Rum is a distilled liquor obtained from the fermentation of molasses and water.

849. *What per cent. of alcohol do the various alcoholic drinks contain?*

Wines, 5 to 25 per cent.; Ciders, 3 to 10 per cent.; Malt liquors, 3 to 13 per cent.; Brandy, 50 to 60 per cent.; Gin, 45 to 55 per cent.; Whisky, 50 to 60 per cent.; and Rum, 60 to 70 per cent.

850. *How are alcoholic drinks adulterated?*

Alcoholic drinks are diluted with large quantities of water to increase the profit in their sale; and then, to restore the strength and flavor lost by the weakening, various vegetable and mineral poisons are added.

851. *What are the principal poisons used to give strength and flavor to alcoholic drinks?*

Cocculus Indicus, Copperas, White Lead, Sugar of Lead, Strychnine, Stramonium, Sulphuric Acid, Nux Vomica, Arsenic, Logwood, Tobacco, Opium, Aloes, Oil of Cloves, etc.

852. *What alcoholic drinks are most adulterated?*

Wines are more adulterated than any other alcoholic drinks; much of what is sold as *wine* is not made from grapes, but is corn whisky after various processes of " doctoring."

853. *Is alcohol a food?*

Alcohol is not a food, since it contains no nutritious elements, and destroys instead of building up the tissues.

854. *Does alcohol slake thirst?*

Alcohol never slakes thirst; it produces thirst by absorbing the water from the tissues and fluids of the body.

855. *What causes the constantly growing thirst of beer drinkers?*

Notwithstanding the fact that beer, on account of the water that it contains, seems to allay thirst, its alcohol, salt, and poisonous adulterations increase thirst and create a craving for stronger drinks.

856. *Are beer drinkers healthy?*

Beer drinkers appear fleshy and robust, but they are really a very unhealthy class of persons. Their fleshy appearance is the result of a species of fatty degeneration; and injuries which would be but slight to other people prove serious to them.

857. *Is new cider a safe and wholesome drink?*

All cider, from that just pressed from the crushed apples to that commonly called hard cider, contains some alcohol; and alcohol in all quantities and combinations is a deadly poison.

858. *Is alcohol a medicine?*

Alcohol has medicinal properties, but they are so overbalanced by the evils resulting from its prescription, that thinking physicians are rapidly abandoning its use for remedies which are equally good and far safer.

859. *Are not some forms of alcoholic drinks allowable luxuries?*

A proper study of the history of drunkard-making will convince any one that *total abstinence from all forms of alcoholic beverages is the only safe course* FOR ALL MEN.

860. *Is alcohol a poison?*

Alcohol is a virulent poison, producing, in large doses, instant death, and in smaller doses, diluted with water, all the effects of slow poisoning.

861. *What is Alcoholism?*

Alcoholism is poisoning by alcohol, and includes all morbid phenomena resulting from the use of alcoholic beverages, from simple drunkenness to the most violent delirium tremens.

862. *What are the two forms of alcoholism?*

Chronic and Acute.

863. *What is Chronic Alcoholism?*

Chronic Alcoholism is a diseased state of the entire man, physical, intellectual and moral, resulting from the habitual use of alcoholic drinks.

864. *What is Acute Alcoholism?*

Acute Alcoholism is simple poisoning, resulting immediately from taking into the stomach a greater or less quantity of the poison; it varies from the slightest tipsiness to profound intoxication and death.

865. *What are the two classes of causes of alcoholism?*

Predisposing Influences and Exciting Causes.

866. *What are the three classes of Predisposing Influences?*

Moral, Social, and Personal Conditions.

867. *What are the principal predisposing moral conditions?*

Unwholesome public sentiment, bad example, and unhappy domestic relations.

868. *In what two grades of society is an unwholesome public sentiment a predisposing influence in alcoholism?*

In communities characterized by poverty and its attendant evils, ignorance and vice, there is but little restraint against the taking of alcoholic poisons; and, on the other hand, in communities characterized by wealth and luxury, there is too often an influence in favor of the use of wine, which brings death by poisoning to many a door.

869. *How does bad example lead to alcoholism?*

The son of the drunkard, surrounded by drinking companions, yields readily to the use of intoxicants; and the young man moving in refined circles is influenced by the example of some " moderate-drinking " doctor or lawyer.

870. *How do unhappy domestic relations lead to alcoholism?*

Unhappy homes and the resulting discontent and lack of motive lead to alcohol poisoning as a means of social entertainment or temporary forgetfulness.

871. *What social conditions predispose to alcoholism?*

Certain occupations, as well as the lack of occupation, lead to alcoholism.

872. *What two classes of occupation predispose to alcoholism?*

Those which bring the workers into direct contact with the poison, as in manufacturing and selling; and those that expose persons to the inclemency of the weather, or to the strain of violent or monotonous toil, as cab driving, iron puddling, type setting, etc.

873. *How does lack of occupation lead to the use of alcoholic drinks?*

Club members, as well as common loafers, while away the time in " treating " and being treated; and the poor unfortunate laborer, who fails to find employment, resorts to drink as a means of drowning his sorrow.

874. *What personal conditions predispose to alcoholism?*

Hereditary taint, chronic disease, prescription by physicians, use of tobacco, etc.

875. *What effect has heredity upon alcoholism?*

Alcoholism in the parent transmits its taint to the child, and is a strong influence in leading the child in the same path.

876. *How do chronic diseases lead to alcoholism?*

Alcohol is resorted to for temporary relief from weakness and suffering in some forms of chronic diseases.

877. *How do the prescriptions of physicians cause chronic alcoholism?*

Physicians prescribe alcoholic stimulants for convalescent patients, and frequently arouse a hereditary or emotional tendency which terminates in death by alcohol poisoning.

878. *How does the use of tobacco lead to alcoholism?*

Tobacco vitiates the taste, depraves the moral sense, and produces a depressed state and a craving for stimulants which leads to the use of alcoholic beverages.

879. *What is the exciting cause of alcoholism?*

The exciting cause in alcoholism is alcohol, taken into the stomach in some form of fermented beverage.

880. *What two classes of physical effects characterize chronic alcoholism?*

General and Local.

881. *What are the general physical effects of alcohol poisoning?*

The blood is thinned and weakened, unhealthy fatty deposits occur in all parts of the body and all the tissues become impaired and vitiated.

882. *How does alcohol act upon the tissues of the body?*

Alcohol hardens the tissues by absorbing water from them.

883. *What effect has alcohol on the coats of the stomach?*

A small amount of alcohol taken into the stomach irritates its delicate mucous coat and produces dilation and congestion of its blood capillaries; repeated doses harden the tissues, make the congestion permanent, and finally cause inflammation and ulceration.

884. *What effect has alcohol on the liver?*

The effects of alcohol upon the liver are similar to those upon the stomach; congestion, inflammation, and ulceration. Fatty degeneration of the liver is a common result of alcohol poisoning.

885. *How does alcohol affect the lungs?*

Congestion of the lungs and pneumonia are frequently caused by the use of alcoholic stimulants.

886. *What effect has alcohol upon the brain?*

Alcohol produces inflammation and hardening of the brain substance.

887. *How are functional derangements produced by chronic alcohol poisoning?*

The various forms of functional derangements resulting from

chronic alcohol poisoning are due to the effects of the poison on the nerve centers.

888. *What general functional derangements characterize chronic alcoholism?*

General sensibility is lessened; disordered motion is produced, beginning with "slight unsteadiness of nerves," and frequently terminating in true paralysis.

889. *How does alcohol poisoning affect the special senses?*

Alcohol deadens the organs of special sense and lessens the intensity and accuracy of their perceptions.

890. *What is the character of the psychical derangements produced by chronic alcoholism?*

Alcohol has a debasing influence upon the mind, which grows with its use, involving in turn the moral sense, the will, and the intellect.

891. *How does chronic alcoholism affect the moral sense?*

The sense of moral obligation is gradually destroyed; sentiments of honor, reputation, decency, and affection for the family give place to dishonesty, indifference, vulgarity, and selfishness.

892. *How does chronic alcoholism affect the will?*

The will is rapidly enfeebled by the habitual use of alcoholic drinks, until the drinker becomes utterly vacillating and irresponsible.

893. *How does chronic alcoholism affect the intellect?*

The progressive mental deterioration, resulting from the use of alcoholic drinks, finally reaches the intellect; slowness in thought, difficulty in fixing the attention, loss of memory, and impairment of judgment characterize the confirmed drunkard.

894. *What are the forms of acute alcoholism?*

Acute alcoholism, resulting from drinking at one time a greater or less amount of intoxicating liquors, has three distinct forms: simple intoxication, convulsions, and profound stupor.

895. *What does the word intoxicate mean?*

The word intoxicate means to poison, being derived from the Latin *toxicum*, which means poison.

896. *What are the three stages of intoxication?*

Simple intoxication manifests in turn three well-marked stages: 1st, increased functional activity; 2nd, functional perversion; and 3rd, depression, ending in stupor.

897. *What are the characteristic marks for the diagnosis of alcoholism?*

Simple intoxication is unfortunately too well known to need

rules for its determining; the watery eyes, flushed cheeks, thick utterances, and staggering gait are known to every school-boy and girl as the marks of a drunken person. The red nose, trembling hand, and lack of honor are equally familiar characteristics of the chronic state.

898. *What is the prognosis in alcoholism?*

Simple intoxication is generally followed by rapid recovery, marked by headache and a general feeling of "meanness." The prognosis in chronic alcoholism is always gloomy; few ever recover.

899. *What are the two kinds of treatment for alcoholism?*

Curative and Preventive Treatment.

900. *What are the two phases of curative treatment?*

Curing simple intoxication and curing chronic alcoholism.

901. *How may a drunken person be sobered?*

Recovery from acute alcohol poisoning is usually rapid, owing to the speedy elimination of alcohol from the system; but where more prompt restoration is necessary, vomiting, dashing cold water on the face and neck are usually resorted to.

902. *How is chronic alcoholism cured?*

Since chronic alcoholism results from the habitual use of alcoholic drinks, the essential condition of a cure is a total discontinuance of the use.

903. *What two means are there of securing the abandonment of the use of the poison?*

Drinking is abandoned either by personal resolution and self-control, stimulated by a remnant of honor, and the love of friends, or by deprivation by friends or the state of the personal freedom necessary to obtain the poison.

904. *What are Inebriate Asylums?*

Inebriate Asylums are institutions where drunkards are placed under such restraint as is necessary to prevent their obtaining alcoholic poison in any of its forms, and where the attempt is made to restore both mind and body to their normal condition.

905. *What are the two classes of Inebriate Asylums?*

State and Private.

906. *Has the state the right to deprive the drunkard of his liberty for the purpose of curing him?*

The state has the right to control any person who has become imbecile and vicious, and should pass laws by which drunkards, after an inquest similar to that held to establish lunacy, would be placed in asylums for one or two years.

907. *Is a drunkard a dangerous member of society?*

A confirmed drunkard is not only an immoral man and a criminal in intention if not in act, but he is a corrupter of morals and a breeder of crime. His influence is wholly pernicious, and the state should protect itself by confining and, if possible, curing him.

908. *What is Delirium Tremens?*

Delirium Tremens is a delirious state in which the drunkard sees "snakes," and all kinds of vermin swarming about him ; it usually follows an attempt to "sober up" after a prolonged debauch, but may be induced by prolonged excess itself.

909. *Does alcohol produce insanity?*

Delirium Tremens frequently ends in insanity, and it is asserted upon good medical authority that forty per cent. of all insanity is due directly or indirectly to the use of alcoholic beverages.

910. *What are the two modes of preventive treatment of alcoholism ?*

By inculcating in those not addicted to it a moral opposition to its use, and by the enactment of such prohibitory laws by the state as will make it impossible to obtain it.

911. *How can a healthy moral sentiment against the use of alcohol be secured ?*

By teaching thoroughly in the school-room and the family its true effects in all forms upon both body and mind.

912. *What steps has the state taken to prevent the increase of the disease ?*

Laws have been passed making it a crime to sell or give distilled alcoholic poisons to minors and persons of unsound mind. These laws should include all forms of the poison.

913. *What are Narcotics ?*

Narcotics are nerve poisons which in their earlier effects and in small doses stimulate, but whose final effects are stupifying.

914. *What are the most pernicious narcotic poisons ?*

Tobacco, Opium, Chloral, and Cocaine.

915. *What is Tobacco ?*

Tobacco, *Nicotiana tabacum,* is a sturdy plant from three to six feet in height, whose large oblong leaves are dried and used in various ways to produce toxic self-abuse.

916. *How is tobacco prepared for chewing?*

Tobacco for chewing is rolled into little twists, or pressed into cakes or "plugs," or cut into fine shreds called "fine-cut."

917. *How is tobacco prepared for smoking?*

Tobacco for smoking is rolled into cigars or cut into small particles for use in pipes and in the making of cigarettes.

918. *What is Snuff?*

Snuff is very finely powdered tobacco leaves.

919. *How is Tobacco adulterated?*

Tobacco in all its forms is very largely adulterated: molasses, licorice, glycerine, and various aromatic flavors are used in chewing tobacco; nitrate of potash, opium, salt, etc., in smoking tobacco; and lime, powdered glass, oxide of lead, etc., in snuff.

920. *How is Tobacco used?*

Tobacco is used in three ways; chewing, in which a portion of the leaf is held in the mouth and crushed by the teeth, mingling its filthy and poisonous principles with the saliva, from which they are absorbed into the blood; smoking, in which the smoke of the burning tobacco in cigar, cigarette or pipe is drawn back into the mouth and lungs; and snuffing, in which minute particles are brought into contact with the delicate mucous membrane of the nostril by sudden inhaling.

921. *What is Nicotine.*

Nicotine is the poisonous principle of tobacco, occurring as an oily, colorless, slightly bitter liquid.

922. *What is the amount of Nicotine in tobacco?*

From two to eight per cent. of dried tobacco leaves are nicotine. It is estimated by good authority that the amount of nicotine contained in a single strong cigar, if thrown directly into the circulation, would produce instant death.

923. *What other poisonous substances are found in tobacco?*

Ammonia, causing a dry, parched state of the mouth and throat; carbonic oxide, weakening the action of the heart; and carbonic acid, producing drowsiness and headache.

924. *What are the effects of tobacco upon the tissues of the body?*

Tobacco inflames the tissues of the stomach, lungs, and heart, and thins and vitiates the blood, produces cancerous affections of the throat, and deadens the nerves of the organs of special sense.

925. *What functional derangements result from the use of tobacco?*

Chronic indigestion, palpitation of the heart, uncertainty in the control of the voluntary muscles, and obtuseness of the senses.

926. *What effect has tobacco on the intellectual and moral natures?*

Tobacco blunts the intellect, destroys the memory, and deadens the moral sense.

927. *Does tobacco unfit the users of it for the highest intellectual activities?*

Yes; carefully gathered statistics in our military and public schools show that the users of tobacco cannot maintain the same standing with boys who do not use it.

928. *What effect has tobacco upon the growth and development of the body?*

Tobacco stunts the body as well as the mind; boys who early acquire the habit seldom grow to full stature.

929. *Has tobacco any beneficial results?*

Absolutely none; no right-minded person would teach another not already addicted to the habit to use the filthy poison.

930. *How does the use of tobacco lead to the use of alcoholic drinks?*

Tobacco creates thirst and a craving for stimulants, and at the same time deadens the moral sense, and prepares the way for all forms of self-abuse.

931. *Why is the smoking of cigarettes specially pernicious?*

Cigarettes are usually made from the poorest grades of tobacco, largely adulterated, and in smoking them fragments of paper, tobacco and the drugs used in adulteration are drawn directly into the lungs.

932. *What is "inhaling"?*

Inhaling tobacco is drawing the smoke back into the lungs and allowing it to escape slowly through the nostrils; it is by far the most harmful of all modes of using tobacco, and is a common habit of cigarette smokers.

933. *What steps should the state take to prevent the formation of the tobacco habit?*

The state should, in virtue of its right to prevent the debauching of its citizens, make it a penal offense to sell or give tobacco in any form to minors, and provide for the punishment of those found using it.

934. *How can those who have formed the tobacco habit be cured?*

There is but one cure for the tobacco habit: quit using it at once and forever

935. *How can boys and young men be prevented from forming the tobacco habit?*

Proper instruction on the filthiness and injuriousness of tobacco is a sure safe-guard against its use.

936. *What is Opium.*

Opium is a narcotic drug prepared from the juice of the White Poppy.

937. *What are the most common forms of opium?*

Crude opium is sold in a pasty, solid form, and medicinal preparations are usually in tinctures.

938. *What is Morphine?*

Morphine is an alkaloid salt obtained from opium and possessing all the toxic properties of the crude drug.

939. *How is opium used by those addicted to it?*

Opium, like tobacco, is chewed and smoked; it is also taken in tincture, as Laudanum ; and Morphine is swallowed or injected beneath the skin.

940. *What are the effects of chronic opium poisoning upon the physical organism and its functions?*

Under the use of opium poison general nutrition rapidly fails; the body becomes emaciated and is frequently covered with ulcerous sores; and there is constant and serious nervous derangement.

941. *What effects has opium on the intellectual and moral natures?*

Chronic opium poisoning destroys the intellect. and more effectually than any other of this class of poisons takes away the last vestige of truthfulness and honor. Persons thought to be above reproach in other matters, will lie without scruple concerning their use of the drug.

942. *What is the principal cause of the opium habit?*

The chief cause of the opium habit is the too common prescription of the drug in various forms by physicians, and its indiscriminate use to deaden pain.

943. *How is the opium habit cured?*

There is probably no successful self-treatment for chronic opium poisoning; the only safe course is to place the afflicted one in some properly conducted asylum, and even there the permanent cure is uncertain.

944. *What is Chloral?*

Chloral, or *Chloral Hydrate*, is a powerful sleep-producing drug, obtained from Chlorine, and having in a pure state a close resemblance to crushed alum crystals.

945. *How is chloral used for self-abuse?*

Chloral is used largely among wealthy and refined people, in small and often repeated doses, to produce sleep.

946. *What are the effects of chloral upon the physical organization and its functions?*

Chloral is a treacherous drug; at first it seems to be only a producer of sleep, but it soon involves every tissue of the body in general ruin. Its primary deleterious effects are upon the nervous centers.

947. *What physical derangements are produced by Chloral?*

Chloral produces general intellectual and moral feebleness, exhibited in intellectual dullness, alternating with irritability and peevishness.

948. *How is the chloral habit cured?*

The chloral habit, like the use of opium, is cured simply by quitting the use of the poison, under the guidance and restraint of skilled attendants.

949. *What is Cocaine?*

Cocaine is a highly poisonous alkaloid obtained from the leaves of the Coca plant.

950. *What are the effects of Cocaine poisoning?*

The habitual use of cocaine produces anatomical and functional derangement of the vital centers, inducing a gloomy sort of mania.

951. *How can the cocaine habit be cured?*

The cocaine habit can be successfully treated only in an asylum under the control of a competent specialist.

GENERAL HYGIENE AND THERAPEUTICS.

Didactic Note.—Some hygienic and therapeutic matter is scattered throughout the book in the various divisions; it is the design to gather under this general head a brief outline of the two contrasted states of Health and Disease. Upon this skeleton the intelligent teacher can build a beautiful form of rules for the preservation of health and the prevention and cure of disease. Interesting and attractive as is the study of the human body, "the casket of the soul," this purely intellectual delight must ever be subordinated in the instruction of youth to the utilitarian end. Make all the instruction tend toward the securing of such a knowledge of hygienic principles and the inculcating of such habits of self-control as will establish and maintain a healthy physical organism, controlled by a vigorous, sane mind. Practical hygienic topics can be presented in the school-room in such a manner as to start discussion in the homes, and much good be accomplished as a sort of medical missionary; but do not attempt to make the "hull deestrict" over in a day. Be moderate in your demands of your pupils and modest in presenting the truth.

ANALYTIC OUTLINE.

General Hygiene and Therapeutics.
- 1^1 Health.
  - 1^2 Characteristics.
    - 1^3 Good physique.
    - 2^3 Adaptability to climatic changes.
    - 3^3 Power of Endurance.
    - 4^3 Resistance to morbic influences.
    - 5^3 Self-control.
    - 6^3 Cheerfulness.
  - 2^2 Conditions.
    - 1^3 Proper diet.
    - 2^3 Pure air.
    - 3^3 Bathing.
    - 4^3 Clothing and shelter.
    - 5^3 Exercise and rest.
    - 6^3 Correct habits.
    - 7^3 Pure thoughts.
  - 3^2 Preservation.
    - 1^3 Secure the "conditions."

2³ Correct promptly any bad symptoms.
2¹ Diseases.
  1² Characteristics.
    1³ Structural deterioration.
    2³ Functional derangement.
    3³ Psychical derangement.
  2² Causes.
    1³ Predisposing.
    2³ Exciting.
  3² Classification.
    1³ As to extent.
      1⁴ General.
      2⁴ Local.
    2³ As to character.
      1⁴ Structural.
      2⁴ Functional.

3³ As to infectiousness.
  1⁴ Contagious.
  2⁴ Non-contagious.
4³ As to intensity and duration.
  1⁴ Acute.
  2⁴ Chronic.
5¹ As to mode of progress.
  1⁴ Continuous.
  2⁴ Periodic.
6³ As to distribution.
  1⁴ Sporadic.
  2⁴ Epidemic.
4² Treatment.
  1³ Curative.
  2³ Preventive.

---

952. *What is Health?*

Health is a state of physical, intellectual, and moral soundness, in which the individual enjoys his existence and the performance of all the functions of both body and mind.

953. *What are the characteristics of health?*

Health is characterized by a good physique, power to adapt one's self to climatic changes, power of endurance of both physical and mental strain, power to resist disease, ready self-control, and cheerfulness.

954. *What are the conditions of health?*

Proper diet, exercise and rest, pure air, bathing, clothing and shelter, correct habits, and pure thoughts.

955. *What is "healthy food"?*

Any article of food which contains proper nutriment is healthy for persons whose stomachs will digest it. No rule can be given in this matter.

956. *What is the best rule for selecting food?*

The best rule is to eat the various kinds of food in their seasons, selecting according to taste and former experiences.

957. *What amount of food should be eaten at each meal?*

No precise rules can be given for determining the amount of food to be eaten at one time. Eat what you want is a good general rule; those who are always dieting are truly the most miserable of people.

958. *What effect should occupation have on the amount and kind of food?*

Persons engaged in physical labor need more tissue-building food, and usually a greater quantity of food than those engaged in intellectual pursuits; but in all employments the hardest workers are generally good eaters.

959. *How should food be eaten?*

Food should be eaten at regular intervals, amidst pleasant surroundings, slowly, and with the mirth and relish that come from social conversation; food eaten in mental depression or ill-humor frequently lies for several hours in the stomach undigested.

960. *What is Hunger?*

Hunger is a craving for food, caused by prolonged abstinence, and characterized by a feeling of uneasiness, faintness and pain.

961. *What is Thirst?*

Thirst is an uneasy, somewhat painful sensation, indicative of a lack of water in the system.

962. *Is there danger in drinking water when heated by violent exercise?*

A reasonable amount of pure, moderately cool water will not ordinarily injure a person heated by exercise, no matter how warm he may be.

963. *Should water be drank "during meals"?*

A glass of pure, cool water drank during a meal aids rather than retards digestion; but a large quantity of water taken into the stomach dilutes the digestive fluids and weakens the digestion.

964. *Why is pure air necessary to good health?*

Air furnishes the oxygen for the blood, and upon the deep, full breathing of pure air, more perhaps than upon any other one thing, depends robust, perfect health.

965. *How does foul air produce disease?*

Foul air brings microscopic germs from sewage and garbage into contact with tissues in which they germinate and breed disease.

966. *What are the evil effects of breathing air in crowded rooms?*

Expired air is heavily charged with carbonic acid, which is

rebreathed in crowded rooms, producing drowsiness and head-ache.

967. *What is the true theory of ventilation ?*

The object to be sought in ventilation is to supply pure air without subjecting those for whom it is intended to the evil effects of a draft; it is far better to rebreathe air tainted with car-bonic acid than to suffer from a draft of cold air.

968. *What is the best ventilation for closed rooms in winter ?*

The best ventilation is that furnished by a strong entering current of pure warm air from a hot-air furnace.

969. *How often should the body be bathed ?*

No rule can be given; the body should be kept clean from the accumulations of excrementitious substances and the dust which gathers upon its damp surface.

970. *Should a person bathe in cold or in warm water ?*

Water for bathing the body should not be so cold as to pro-duce chilliness, nor so warm as to weaken and enervate.

971. *What clothing and shelter does health require ?*

The maintenance of health requires such clothing and shelter as will furnish protection against the inclemency of the weather.

972. *What is the proper degree of heat for the school-room ?*

A school-room in which children are sitting without physical exercise should be kept at a temperature of 70° or 72°. School-rooms are more often too cold than too warm.

973. *Why is exercise necessary to health ?*

Strength and tone in all the textures of the body result from proper exercise; an unused muscle or organ soon becomes inca-pacitated for the performance of its functions.

974. *What amount of exercise is required for good health ?*

Most people, even in sedentary employments, get from their daily avocations sufficient exercise, but when this is not enough, a short, rapid walk with a pleasant, sociable companion is the best form of exercise.

975. *What is rest ?*

Rest is a cessation of functional activity to permit the organs to repair the tissues destroyed by their use; the most perfect rest is in sleep.

976. *How much sleep is required for health ?*

The average amount of sleep required by adults is eight hours; children should have ten hours. As a rule, those following intellectual pursuits require more sleep than those engaged in physical toil.

977. *What effect have habits upon health?*

Good health depends upon right living; and right living is merely a sum of good habits. Bad habits are utterly incompatible with good health.

978. *Is intellectual exercise necessary to good health?*

A certain amount of intellectual exercise is essential to good health; just as "a sound mind requires a sound body," so a healthy physical organism requires an active, controlling mind.

979. *Is purity of thought necessary to health?*

Pure thoughts are necessary to maintain the intellectual tone required for a clean, strong, right-acting body.

980. *What are the requirements for the preservation of health?*

The preservation of the health requires that all the conditions of health be secured, and that prompt action be taken to correct any derangement in structure or functions.

981. *What is Disease?*

Disease is any variation from the normal structure or functions of the body.

982. *What are the characteristics of disease?*

Disease manifests itself in a deterioration of tissues, a lessening or perversion of functions, and a weakening of the intellectual and moral tone.

983. *What are the two classes of causes of diseases?*

Predisposing and Exciting.

984. *What are the Predisposing Causes?*

Predisposing Causes are peculiarities or conditions of the individual which make him susceptible to the action of disease-producing elements.

985. *What are Exciting Causes?*

Exciting Causes are disease germs, malarial poisons, and various forms of irritating substances and forces which, by their direct action, produce disease.

986. *What effect has occupation upon disease?*

Different occupations not only predispose to different diseased states, but frequently become the exciting causes, as painters suffer from lead colic, coachmen have rheumatism in their shoulders, etc.

987. *How are diseases classified?*

Diseases are classified with respect to extent of the part affected, character of the affection, infectiousness, intensity, and duration, mode of progress, and distribution in communities.

988. *What are the two classes of diseases with respect to extent of the part affected?*

General and Local.

989. *What are the two classes of diseases with respect to character of the affection?*

Structural and Functional.

990. *What are the two classes of diseases with respect to infectiousness?*

Contagious and Non-contagious.

991. *What are the two classes of diseases with respect to intensity and duration?*

Acute and Chronic.

992. *What are the two classes of diseases with respect to mode of progress?*

Continuous and Periodical.

993. *What are the two classes of diseases with respect to distribution in communities?*

Sporadic and Epidemic.

994. *What are the two kinds of treatment of diseases?*

Curative and Preventive.

995. *What is curative treatment?*

Curative treatment consists in removing the causes of disease, and aiding nature in restoring healthy structure and functions.

996. *What are medicines?*

Medicines are various substances used in the treatment of disease, to prevent or correct derangement of structure or functions.

997. *What is nursing the sick?*

Nursing is caring for the sick in such a way as to give nature the best possible conditions for restoring health; it is by far the most important part of curative treatment.

998. *What is preventive treatment?*

The preventive treatment of disease consists in caring for the body and mind in such a manner as to maintain perfect health of structure and functions.

999. *What are the functions of the state in the prevention of disease?*

It is the duty of the state to enact such laws as will remove conditious which injure the health of the people; it is in accordance with this that Boards of Health created by law compel the removal of garbage, establish quarantines against disease-infected communities, etc.

1000. *Why should Physiology be studied?*

Human happiness depends upon a healthy condition of the body, and the health of the body depends upon the observance of a few simple *Laws of Health;* therefore the study of Physiology becomes a duty binding upon all who would promote their own happiness and that of others.

HOLBROOK-ROHRER BOOK-KEEPING.

PRICE LIST.

The Following Editions are Published

1st. The Primary Book-keeping comprising a full list of questions and answers on the Science and principles of double entry book-keeping.

A Model Journal showing how to journalize every conceivable business transaction from a memorandum given.

A Model Ledger showing how to post, and close the ledger.

Mercantile Forms, glossary of business terms, tables for computing time and wages.

Easy exercises for students to journalize and post.

A complete new set for general retail store.

A new set for farmers or planters.

These are the same sets used in the Normal University, Lebanon, O.

Price by mail, each - - - **60c.**
Price for first introduction, each - **35c. $4 per doz.**
3 Blank Books for the above - - **75c.**

2nd. The Condensed Counting House and School Edition as used in the Normal University. This edition includes the exercises of the Complete Edition, excepting the worked sets. It has full models for all books, is a very complete book of 160 large octavo pages and is well bound in crimson cloth.

Price by mail, each - - - - **$1.50**
Price for first introduction, each - - - **1.00**

3rd. The Complete Counting House Edition is a large book of 320 octavo pages. It comprises definitions and explanations and worked sets in every kind of single and double entry book-keeping, viz: Merchandizing, wholesale and retail, farming, commission, banking, steamboating, railroading and manufacturing as transacted by individuals, partners, corporations and joint stock companies. Forms for business letters, notes, drafts, checks, account sales; a full treatise on commercial calculations, and additional easy exercises for class practice. This is the most complete work on book-keeping published. Bound in handsome crimson cloth.

Price by mail, each - - - - **$3.00**
Price for first introduction, each - - - **2.00**

4th. Key to the foregoing texts giving the journal entries and instructions for each set. By the aid of this book the teacher will be perfectly safe in conducting a class through the most difficult sets in the Counting House edition. Bound in cloth.

Price by mail, - - - - - **$2.00**

5th. Holbrook's Book-keeping Blanks as used in the University at Lebanon, O., embraces the following books: Day Book, Journal, Six or Special Column Journal, Cash-book, Bill-books, Ledger, and all the books necessary for the Counting House edition, all in 2 strongly bound books.

Price for the 2 Volumes - - - **$1.25**

6th. A lecture on the science of keeping accounts, showing its importance, difficulties, methods with illustrations. This lecture makes a good outline or basis for a series of lectures that can be given by the teacher to a school or miscellaneous audience. 32 large pages.

Price by mail, - - - - - **25c.**

C. K. HAMILTON & CO., Publishers, Lebanon, O.

www.ingramcontent.com/pod-product-compliance
Lightning Source LLC
Chambersburg PA
CBHW021811190326
41518CB00007B/546